T0224401

Lecture Notes in Computer Science   8670

Commenced Publication in 1973
Founding and Former Series Editors:
Gerhard Goos, Juris Hartmanis, and Jan van Leeuwen

Editorial Board

More information about this series at http://www.springer.com/series/8851

Ngoc Thanh Nguyen · Ryszard Kowalczyk
Juan Manuel Corchado · Javier Bajo (Eds.)

Transactions on Computational Collective Intelligence XV

 Springer

Editor-in-Chief

Ngoc Thanh Nguyen
Institute of Informatics
Wroclaw University of Technology
Wroclaw
Poland

Co-editor-in-Chief

Ryszard Kowalczyk
School of Software and Electrical
    Engineering
Swinburne University of Technology
Hawthorn
Australia

Guest Editors

Juan Manuel Corchado
Departamento de Infomática y Automática
Universidad de Salamanca
Salamanca
Spain

Javier Bajo
Departamento de Inteligenca Artificial
Universidad Politécnica de Madrid
Madrid
Spain

ISSN 0302-9743
ISBN 978-3-662-44749-9
DOI 10.1007/978-3-662-44750-5

ISSN 1611-3349   (electronic)
ISBN 978-3-662-44750-5   (eBook)

Library of Congress Control Number: 2014948788

Springer Heidelberg New York Dordrecht London

Printed on acid-free paper

Springer is part of Springer Science+Business Media (www.springer.com)

Preface

This volume of TCCI is a special issue dedicated to the International Conference on Practical Applications on Agents and Multi-Agent Systems (PAAMS 2012 and PAAMS 2013) held in Salamanca during March 28–30, 2012 and May 22–24, 2013. PAAMS provides an international forum to present and discuss the latest scientific developments and their effective applications, to assess the impact of the approach, and to facilitate technology transfer. PAAMS started as a local initiative, but has since grown to become the international yearly platform to present, to discuss, and to disseminate the latest developments and the most important outcomes related to real-world applications. It provides a unique opportunity to bring multidisciplinary experts, academics, and practitioners together to exchange their experience in the development and deployment of agents and multi-agent systems. PAAMS intends to bring together researchers and developers from industry and the academic world to report on the latest scientific and technical advances in the application of multi-agent systems, to discuss and debate the major issues, and to showcase the latest systems using agent-based technology. This will promote a forum for discussion on how agent-based techniques, methods, and tools help system designers to accomplish mapping between available agent technology and application needs. Other stakeholders will be rewarded with a better understanding of the potential and challenges to the agent-oriented approach. The conference is organized by the Bioinformatics, Intelligent System and Educational Technology Research Group (http://bisite.usal.es/) of the University of Salamanca.

This volume includes the best papers presented at the conference, which were subsequently extended and selected after the peer-review process. In the first paper, Lamarche-Perrin et al. present measures inherited from information theory to evaluate abstractions of large-scale MAS and provide experts with feedback regarding the quality of generated representations. The design and debugging of large-scale MAS require abstraction tools to work at a macroscopic level of description. Agent aggregation provides such abstractions by reducing the complexity of the microscopic representation. Since it leads to information loss, such a key process may be extremely harmful for analysis if poorly executed. In this paper, several evaluation techniques are applied to spatial and temporal aggregation of an agent-based model of international relations. The information from online newspapers constitutes a complex microscopic representation of agent states. Lamarche-Perrin et al.'s approach is able to evaluate geographical abstractions used by domain experts to provide efficient and meaningful macroscopic representations of the world global state in space and in time.

Alexei Sharpanskykh and Kashif Zia, in the second paper, discuss and investigate the role of emotions in social decision-making in large technically assisted crowds. For this a formal, computational model is proposed, which integrates existing neurological and cognitive theories of affective decision-making. Based on this model, several variants of a large-scale crowd evacuation scenario were simulated. By analysis of simulation results, it was established that (1) human agents supported by personal assistant devices are recognized as leaders in groups emerging in evacuation; (2) spread of emotions in a crowd increases the resistance of agent groups to opinion changes; (3) spread of emotions in a group increases its cohesiveness; and (4) emotional influences in and between groups are, however, attenuated by personal assistant devices when their number is large.

In the third paper, Ksontini et al. propose to improve the validity of traffic simulations in the (sub-)urban context, with better consideration of driver behavior in terms of anticipation of positioning on the lanes and occupation of space. They introduce a model based on a multi-agent approach and the emergence concept. This model considers that each driver perceives the situation in an ego-centered way and readapts the road space using the virtual lane concept. They implement the model with the traffic simulation tool ArchiSim. The so obtained simulator intends to reproduce the observed behavior such as filtering between vehicles (two-wheels and emergency vehicles), repositioning on lanes when approaching the road intersections, and "exceptional" situations (stranded vehicle or improperly parked, etc.).

In the fourth paper, Philippe Mathieu and Yann Secq show how to leverage information from the order books such as the best limits, the bid-ask spread, or waiting cash to adapt more effectively to market offerings. Like B. Arthur, they use learning classifier systems and show how to adapt them to a multi-agent system. In the study of financial phenomena, multi-agent market order-driven simulators are tools that can effectively test different economic assumptions. Many studies have focused on the analysis of adaptive learning agents carrying on prices. But the prices are a consequence of the matching orders. Reasoning about orders should help to anticipate future prices. While it is easy to populate these virtual worlds with agents analyzing "simple" prices shapes (rising or falling, moving averages, etc.), it is nevertheless necessary to study the phenomena of rationality and influence between agents, which requires the use of adaptive agents that can learn from their environment. Several authors have obviously already used adaptive techniques but mainly by taking into account prices historical. But prices are only consequences of orders, thus reasoning about orders should provide a step ahead in the deductive process.

In the fifth paper, Li et al. target the coupling similarities from these three perspectives and design a novel classification method that applies a weighted K-nearest centroid to obtain the coupled similarity for non-iid data. From value and attribute perspectives, coupled similarity serves as a metric for nominal objects, which consider not only intra-coupled similarity within an attribute but also inter-coupled similarity between attributes. From the object perspective, they propose a more effective method that measures the centroid object by connecting all related objects. Extensive experiments on UCI and student datasets reveal that the proposed method outperforms classical methods for higher accuracy, especially in imbalanced data.

Zied Sellami and Valerie Camps, in the sixth paper, present DYNAMO-MAS, an adaptive multi-agent system that automates these tasks by co-constructing an ontology from texts with an ontologist. Terms and concepts of a given domain are agentified and they act, according to the adaptive multi-agent system (AMAS) approach, by solving the noncooperative situations they locally perceive at runtime. These agents cooperate to determine their position in the AMAS (that is, the ontology) thanks to (i) lexical relations between terms, (ii) some adaptive mechanisms enabling addition, removing, or moving of new terms, concepts, and relations in the ontology as well as (iii) feedbacks from the ontologist about the propositions given by the AMAS. The paper focuses on the instantiation of the AMAS approach to this difficult problem. It presents the architecture of DYNAMO-MAS, and details the cooperative behaviors of the two types of defined agents for ontology evolution. Evaluations made on three different ontologies are also given to prove that our proposed solution is generic.

Stiborek et al., in the seventh paper, present a self-adaptation mechanism for network intrusion detection system based on the use of game-theoretical formalism. The key innovation of our method is a secure runtime definition and solution of the game and real-time use of game solutions for immediate system reconfiguration. Their approach is suited for realistic environments, where we typically lack any ground-truth information regarding traffic legitimacy/maliciousness and where the significant portion of system inputs may be shaped by the attacker to render the system ineffective. Therefore, they rely on the concept of challenge insertion: we inject a small sample of simulated attacks into the unknown traffic and use the system response to these attacks to define the game structure and utility functions. This approach is also advantageous from the security perspective, as manipulation of the adaptive process by the attacker is far more difficult.

In the last paper, De la Prieta et al. discuss how cloud computing has gained importance at a remarkable pace. The key characteristic of this technology is the possibility to provide new resources to services in an elastic way according to current demand. In contrast to cloud computing, multi-agent systems are the focus on other features such as autonomy, decentralization, auto-organization, etc. De la Prieta et al. demonstrate that these features of MAS are suitable to manage the physical infrastructure of a cloud computing environment; in other words, they present +Cloud which is a cloud platform managed by a multi-agent system.

We thank all the contributing authors, as well as the members of the Program Committee and the Organizing Committee, for their hard and valuable work. Their work has helped to contribute to the success of this symposium. Finally, the iHAS project is acknowledged. We hope the reader will share our joy and find this special issue useful.

This work has been carried out by the project Sociedades Humano-Agente: Inmersion, Adaptacion y Simulacion (iHAS)—TIN2012-36586-C03-03; Ministerio de Economía y Competitividad (Spain); and Fondos Feder.

June 2014   Juan Manuel Corchado
                                                                    Javier Bajo

Transactions on Computational Collective Intelligence

This Springer journal focuses on research on the applications of computer-based methods of computational collective intelligence (CCI) and their applications in a wide range of fields such as the Semantic Web, social networks, and multi-agent systems. It aims to provide a forum for the presentation of scientific research and technological achievements accomplished by the international community.

The topics addressed by this journal include all solutions to real-life problems for which it is necessary to use CCI technologies to achieve effective results. The emphasis of the papers is on novel and original research and technological advancements. Special features on specific topics are welcome.

Edward Szczerbicki   University of Newcastle, Australia
Tadeusz Szuba   AGH University of Science and Technology,
                                Poland
Kristinn R. Thorisson   Reykjavik University, Iceland
Gloria Phillips-Wren   Loyola University Maryland, USA
Sławomir Zadrożny   Institute of Research Systems, PAS, Poland
Bernadetta Maleszka   Assistant Editor, Wroclaw University
                                of Technology, Poland

Contents

Building Optimal Macroscopic Representations of Complex Multi-agent Systems

Application to the Spatial and Temporal Analysis of International Relations Through News Aggregation

Robin Lamarche-Perrin[1]([✉]), Yves Demazeau[2], and Jean-Marc Vincent[1]

[1] LIG, Université Grenoble Alpes, 38000 Grenoble, France
{Robin.Lamarche-Perrin,Jean-Marc.Vincent}@imag.fr
[2] LIG, CNRS, 38000 Grenoble, France
Yves.Demazeau@imag.fr

Abstract. The design and the debugging of large-scale MAS require abstraction tools in order to work at a macroscopic level of description. Agent aggregation provides such abstractions by reducing the complexity of the system's microscopic representation. Since it leads to an information loss, such a key process may be extremely harmful for the analysis if poorly executed. This paper presents measures inherited from information theory to evaluate abstractions and to provide the experts with feedback regarding the quality of generated representations. Several evaluation techniques are applied to the spatial and temporal aggregation of an agent-based model of international relations. The information from on-line newspapers constitutes a complex microscopic representation of the agent states. Our approach is able to evaluate geographical abstractions used by the domain experts in order to provide efficient and meaningful macroscopic representations of the world global state.

Keywords: Large-scale MAS · Agent aggregation · Macroscopic representation · Information theory · Geographical and news analysis

1 Introduction

Because of their increasing size, complexity, and concurrency, current multi-agent systems (MAS) can no longer be understood from a microscopic point of view. Design, debugging and optimization of such large-scale distributed applications require tools that proceed at a higher representational level by providing insightful abstractions regarding the system's dynamics. Among abstraction techniques (*e.g.*, dimension reduction, subsetting, segmentation, clustering [1]), this paper focuses on *data aggregation*. It consists in losing some information regarding the agent level to build simpler – yet meaningful – macroscopic representations.

Such a process is not trivial for the interpretation of the data by the observer. In particular, unsound aggregations may lead to critical misrepresentations of the

© Springer-Verlag Berlin Heidelberg 2014
R. Kowalczyk et al. (Eds.): TCCI XV, LNCS 8670, pp. 1–27, 2014.
DOI: 10.1007/978-3-662-44750-5_1

MAS behavior. Hence, one needs to determine what are the *good* abstractions and how to properly use them. At each stage of MAS development, aggregation processes should be carefully monitored and feedback should be provided regarding the quality of the generated macroscopic representations. A simple example can demonstrate how critical the aggregation process can be. Figure 1 shows two groups of agents that are simplified by two abstract entities with an averaged behavior. Intuitively, group A constitutes a *good* abstraction since the induced global behavior is relatively similar to the microscopic one, unlike group B. Hence, in order to scale-up, aggregation of redundant information should be encouraged to reduce the representation complexity (group A), but details regarding heterogeneous behaviors should be preserved in order to control the information loss and proceed to a sound analysis (group B).

Very little work has been done in the MAS community to quantify such aggregation properties. The main contribution of this paper consists in introducing measures from information theory (Kullback-Leibler (KL) divergence [2] and Shannon entropy [3]) to clarify the notion of *good* aggregation. From these measures, we provide generic feedback techniques and an algorithm that builds multiresolution representations out of hierarchically organized MAS. These techniques and algorithms are applied to the agent-based modeling of international relations: agents represent countries, and their behavior is extracted from on-line newspapers. Geographers exploit multilevel aggregates to build statistics regarding world areas. We show how these geographical abstractions should be used to better understand the system states and its evolution through time.

Section 2 presents the work related to the main concern of this article. Section 3 presents the agent-based model of the ANR CORPUS GEOMEDIA application. Sections 4 and 5 introduce KL divergence and the size of representations to respectively estimate *information loss* and *complexity reduction*. Section 6 shows how these measures can be combined to identify *optimal* aggregations and to build multiresolution representations of hierarchically organized MAS. Section 7 applies

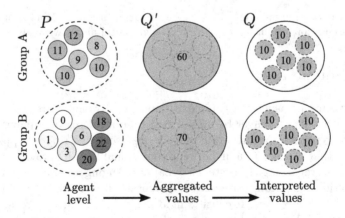

Fig. 1. Averaging the behavior of groups of agents may reduce the redundant information (group A) or lead to an undesired information loss (group B)

these aggregation techniques to the time dimension in order to provide macroscopic representations of the system's dynamics.

2 Related Work

Aggregation can take place in every stage of the MAS development: from its design to its use. Even if abstraction techniques may differ, each stage should carefully take into consideration the quality of the provided aggregations. First, from a software perspective, this section shows that very few research efforts have been done to tackle this issue. (1) Most classical simulation platforms and monitoring systems do not even provide the user with abstraction tools; (2) some do handle the issue, but are still at an early stage of thought. Secondly, on a theoretical aspect, this section explains why classical techniques (*e.g.* data clustering, graph analysis) are not entirely satisfying to build consistent abstractions.

In a comprehensive survey of agent-based simulation platforms [4], Railsback *et al.* evaluate some tools by testing classical features of MAS modeling and analysis. Unfortunately, the abstraction problem is not tackled by this survey, thus indicating that such considerations are seldom if ever taken into account. Most platforms (Java Swarm, Repast, MASON, NetLogo and Objective-C Swarm) are limited to the microscopic simulation of agents. Railsback warns about the lack of "a complete tool for statistical output" in these platforms [4]. The provision of global views on the MAS macroscopic behavior thus constitutes an on-going research topic. Some tools for large-scale MAS monitoring however address this issue. For example, in some debugging systems, abstractions are used to reduce the information complexity of execution traces; however, they are either limited to the simplification of agents internal behavior, and do not tackled multi-agent organizational patterns [5], or they are provided without any feedback regarding their quality for the analysis [6,7].

Some techniques from graph analysis and data clustering build groups of agents out of their *microscopic properties* (see for example [8–10]). Such considerations may meet ours from a theoretical point of view, but the approach presented in this report supports a very different philosophy: *abstractions should be built regarding some macroscopic semantics*. We claim that, to be meaningful, the aggregation process needs to rely on exogenous high-level abstractions defined by the experts. Hence, our approach should rather be related to studies on multilevel agent-based models [11]. These works openly tackle the abstraction problem by designing MAS at several organizational levels according to the expert definitions. Such approaches aim at reducing the computational cost of simulations depending on the expected level of detail. The algorithm and measures presented in this paper may provide a formal and quantitative framework to such researches.

To conclude, aggregation techniques should be more systematically implemented on MAS platforms in order to handle large-scale systems. They should combine consistent macroscopic semantics from the experts and feedback regarding the abstractions quality. In this paper, abstractions used by geographers are evaluated according to their information content.

3   Agent-Based Modeling of International Relations

This section presents the GEOMEDIA agent-based model. It consists in the microscopic representation of countries by *agents* and the macroscopic representation of world geographical areas by *groups of agents* and by *organizations*.

3.1   Microscopic Data: The Agent Level

Let A be a set of agents constituting the MAS microscopic level. Visualization tools aim at displaying and explaining the properties of these agents: their behavior and internal states, the events they are associated with, the messages they exchange, and so on. Given a variable v that expresses such properties, the set of values $\{v(a)\}_{a \in A}$ constitutes the *microscopic representation* of the system (illustrated by distribution P in Fig. 1).

The ANR CORPUS GEOMEDIA project[1] is interested in the analysis of world international relations through a media point of view. This project is conducted in collaboration with geographers and media experts from the CIST (*Collège International des Sciences du Territoire*, Paris). In that context, we make the assumption that citations or co-citations of countries, within news, are good indicators to represent and understand their political, economical and cultural relations. For example, we may assume that an often-cited country is likely to politically interact with the newspaper country. Our agent-based model has two dimensions:

- The agents of the model represent the $|A| = 193$ United Nation member states, selected by geographers depending on their significance for the analysis of international relations.
- The temporal dimension contains $|T| = 90$ weeks, from the 3rd of May 2011 to the 20th of January 2013. This preliminary aggregation to the week level aims at reducing the chaotic variations of the day level and focusing on the more significant variations related to media events.

The experiments presented in this paper focus on a very basic variable: the number of articles that cite a country during a given time period. We use the 59,234 articles published by the "world" RSS flow of *The Guardian*[2] during the analyzed period and stored in the GEOMEDIA database. For each article, we look for the occurrences of the country names, the country adjectives, and the inhabitants names (*e.g.*, "Spain", "Spanish", and "Spaniard(s)" for the Spain agent). Thus, for each agent a and time period t, we count the number of articles $v(a,t)$ that "cite a during t". A total of 138,811 citations have been found within the dataset, distributed within 77 % of the articles (3 citations/article in average if we set aside the 23 % that contain no citation at all).

---

[1] Founded by the French *National Agency for Research* (ANR-GUI-AAP-04). See the dedicated website for details: http://geomedia.hypotheses.org/.
[2] http://www.theguardian.com/world

In order to spot critical aspects of the international systems, geographers are interested in detecting significant events in the news. Such events correspond to unexpected values of the variable according to the following hypothesis: *the citation numbers of countries are homogeneous through time*. In that sense, the marginal values of the dataset give the expected citation number. For an agent a and a time period t, we thus expect the observed value $v(a,t)$ to be close to:

$$v^*(a,t) = \frac{v(a,T)\ v(A,t)}{v(A,T)}$$

where $v(a,T)$ is the citation number of agent a during the whole observation period T; $v(A,t)$ is the total number of citations, regarding all agents in A, during the time period t; and $v(A,T)$ is the total number of citations within the dataset. A media event thus correspond to a high observed value $v(a,t)$ compared to the expected value $v^*(a,t)$.

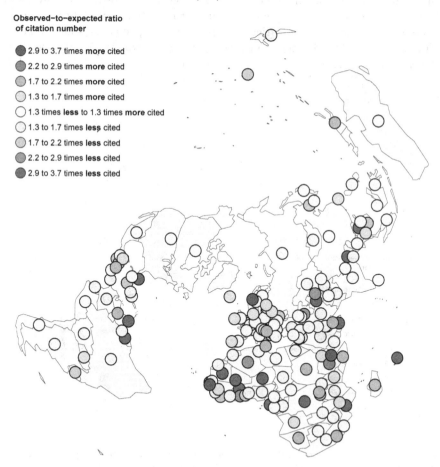

Fig. 2. Observed-to-expected ratio of countries citation within the articles published by *The Guardian* during the month of July 2011 (red circles indicate media events) (Color figure online)

Figure 2 above displays the *observed-to-expected ratio of citation number* $v(a,t)/v^*(a,t)$ for each country $a \in A$ during $t =$ "the month of July 2011". A detailed survey of this map allows to identify geographical areas that have been unexpectedly over-cited during this time period: at the national level (*e.g.*, Norway, Djibouti, Guinea-Bissau) and at higher levels, *i.e.* for groups of countries (*e.g.*, Europe, Horn of Africa). However, the quantity of information displayed in such a microscopic representation makes it quite hard to read. In particular, the visual clutter in dense areas prevents the proper interpretation of data. To overcome this difficulty, Figs. 3 and 4 propose to focus on areas of particular interest (resp. Europe and Africa). In the following sections, we will focus on two particular events that occurred in these geographical areas:

1. The observed citation number of the `Norway` agent is 3.7 times higher than expected (see Fig. 3). This is explained by the terrorist attacks that occurred in Norway the 22th of July 2011[3]. This event belongs to the national level and thus constitutes a *microscopic event* within the system's spatial dimension.
2. Countries of the Horn of Africa also present unexpected citation numbers (from 1.9 times to 3.4 times the expected value for `Rwanda`, `Sudan`, `Somalia`, `Ethiopia` and `Djibouti`, see Fig. 4). This is explained by the food crisis that

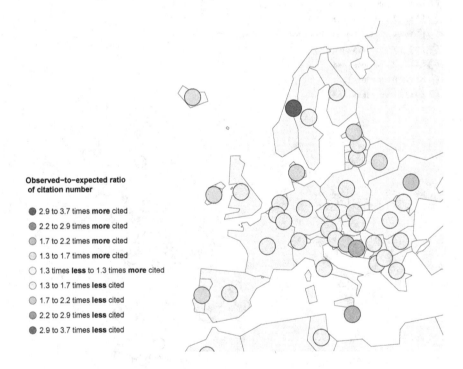

Observed–to–expected ratio of citation number

- 2.9 to 3.7 times **more** cited
- 2.2 to 2.9 times **more** cited
- 1.7 to 2.2 times **more** cited
- 1.3 to 1.7 times **more** cited
- 1.3 times **less** to 1.3 times **more** cited
- 1.3 to 1.7 times **less** cited
- 1.7 to 2.2 times **less** cited
- 2.2 to 2.9 times **less** cited
- 2.9 to 3.7 times **less** cited

Fig. 3. Observed-to-expected ratio of citation number (zoom on European countries with a national event detected in the `Norway` agent)

---

[3] http://en.wikipedia.org/wiki/2011_Norway_attacks

Fig. 4. Observed-to-expected ratio of citation number (zoom on African countries with an international event detected in the `Horn of Africa` group of agents)

has been reported in this world area starting from the beginning of July 2011[4]. Unlike the previous one, this event is not located at the national level, but regards a group of agents located in a spatially spread out area.

3.2   Macroscopic Data: Groups and Organizations

Even if the maps in Figs. 3 and 4 allow to easily spot the two events we are interested in, they do not manage to give the global overview of the world-wide system that is necessary for an informed analysis. *Data aggregation* aims at resuming the microscopic information to provide such an overview.

A *group* $G \subset A$ is subset of agents that are members of a consistent organizational pattern. It can be interpreted as an *abstract entity* that sums up the behavior of its underlying agents. Hence, groups satisfy a recursive definition: a group is either an agent or a set of groups. Quantitative variables expressing agents properties may be extended on groups according to an aggregation operator: *e.g.*, sum, mean, median, extrema [1]. In our case, since we work with extensive variables, *i.e.* variables that are proportional to the aggregate size, $v(G, t)$ is defined as the *sum* of the values of the underlying agents (see distribution Q' in Fig. 1):

---

[4] http://en.wikipedia.org/wiki/2011_East_Africa_drought

$$v(G,t) = \sum_{a \in G} v(a,t)$$

We define an *organization* O as a set of groups that constitutes a *partition* of the agent set A. Thus, in the scope of this paper, each agent is always a member of one and only one group. The set of group values $\{v(G,t)\}_{G \in O}$ composes a *macroscopic representation* of the system with respect to a given organization. It simplifies the variable distribution, from the detailed microscopic representation (distribution P in Fig. 1) to an aggregated one (distribution Q').

Fig. 5. Observed-to-expected ratio of citation number for the groups of agents defined by the 3rd level of the WUTS hierarchical organization

In order to be consistent with the observer's background knowledge, groups and organizations should be derived from the structural and semantical properties of the agent space. In our context, the world's *social*, *political*, and *economic* organization is used by geographers to represent and explain the data. Moreover, in this paper, we also focus on the world's *topological* organization in order to be consistent with classic geographical representations. Groups thus aggregate adjacent territories that share a cultural and historical background. In the following experiments, we consider two hierarchical organizations of countries that meet these needs, namely WUTS [12] and UNEP [13]. Such organizations define multilevel nested groups commonly used by geographers to build global statistics regarding world areas, from the microscopic level of agents to the full aggregation (see [12] for a detailed presentation of these multiscale organizations).

As an example, Fig. 5 provides with the observed-to-expected ratio of citation numbers aggregated according to WUTS_3, the 3rd level of the WUTS hierarchy. Because of the data reduction, this map is much easier to analyze than the microscopic one (see Fig. 2). In particular, the food crisis that occurred in the Horn of Africa group of agents is resumed by one observed macroscopic value that is globally 1.8 times higher than the expected citation number. In that case, the aggregation macroscopically represents the corresponding event. However, most of the microscopic variations have been suppressed by the aggregation process: for example, the events that occurred in the Norway agent are no longer represented. We thus need to control the aggregation process in order to visualize events at different levels depending on their spatial granularity. The following sections present an aggregation technique to automatically build such multiresolution representations of MAS.

4   KL Divergence as a Measure of Organization Quality

When an observer tries to interpret the data that is contained in a macroscopic representation, she necessarily makes an assumption regarding the distribution of the aggregated values over the underlying agents. For example, in Fig. 1, the observer considers that each agent has the same weight in the group. It is thus underlined that aggregated values are *uniformly distributed* over the agents (from Q' to Q). Consequently, some groups are more suitable than others to summarize the microscopic information: using group A seems relevant since P is close to Q, unlike group B. Hence, organizations should be carefully chosen in order to provide accurate abstractions. In particular, they should only aggregate homogeneous and redundant distributions of the displayed variable.

Among classical similarity measures to compare a source distribution P with a model distribution Q, Kullback-Leibler (KL) divergence is of highest interest because of its interpretation in terms of information content. This section shows how it can be exploited to provide feedback regarding the quality of groups and organizations and to ensure their proper interpretation by the observer.

4.1   Formalization and Semantics of KL Divergence

Formally, KL divergence measures the number of bits of information that one loses by using the model distribution Q to find the optimal binary coding of countries associated to articles, instead of using the source distribution P [2]. In other words, KL divergence estimates the information that is lost by the aggregation process. But more generally, it is a measure of dissimilarity between two probability distributions. Hence, it can be interpreted as a *fitness function* between a source P and a model Q.

In our case, the "uniform hypothesis" is not suitable to interpret an aggregated representation. Indeed, for a given group, countries *do not have* the same weight regarding citation number. For example, the observer may assume that, within the Northern America group, the USA agent usually accumulates much more citations than the Canada and the Mexico agents. The aggregated value should thus be interpreted depending on that fact. The marginal values can be used to interpret an aggregated representation Q' and to give the corresponding model Q: the citations associated to a group of agents during a time period are distributed according to the total citation numbers of the underlying agents over the whole dataset. Given an agent a in a group G and a time period t, the interpreted citation number is thus given by the following formula:

$$Q(a,t) = v(G,t) \, \frac{v(a,T)}{v(G,T)}$$

This interpreted value is then compared to the observed microscopic value: $P(a,t) = v(a,t)$. From the KL formula in [2], we define the *divergence* of a group G (or *information loss*, in bits) as follows:

$$\text{loss}(G,t) = \sum_{a \in G} P(a,t) \log_2 \left(\frac{P(a,t)}{Q(a,t)} \right)$$

$$= \sum_{a \in G} v(a,t) \log_2 \left(\frac{v(a,t) \, v(G,T)}{v(G,t) \, v(a,T)} \right)$$

As we assume that aggregated values are thus distributed among underlying agents, a group whose internal distribution is very close to the observed distribution (as group A in Fig. 1) will have a low divergence, and conversely (as group B). Moreover, KL divergence verifies the *sum property* [14], meaning that the divergence of disjoint groups is the sum of their divergences. Therefore, for an organization O, we have:

$$\text{loss}(O,t) = \sum_{G \in O} \text{loss}(G,t)$$

4.2 Divergence is Correlated with the Source of Information

This first experiment aims at showing an essential feature of the aggregation process: its quality depends on the context of the analysis. Figure 6 presents the KL divergence of groups defined by the WUTS_3 macroscopic organization for two different newspapers (*The Guardian* and *The New York Times*) that have been observed during the month of July 2011. The *darker* a group is, the *higher* its

KL divergence is, the more *heterogeneous* its internal distribution is. Such groups should not be used for aggregation since they induce a misleading interpretation of the data. In this case, the real microscopic representation significantly diverges from the macroscopic model, making these groups unsuitable for the analysis. On the contrary, *bright* groups of countries constitutes *good* abstractions in terms of information content. The aggregated representation they provide regarding the corresponding geographical area fits with the microscopic data and can thus be properly interpreted by the observer.

In the case of *The Guardian* (*cf.* Fig. 6(a)), the groups with a high divergence are the location of microscopic events that can be spotted in Fig. 2. In these cases, the suppression of the corresponding microscopic variations induces a significant information loss. Divergence thus indicates heterogeneous behaviors in lower levels that should be detailed in order to reveal significant microscopic events. In the case of the *The New York Times* (*cf.* Fig. 6(b)), the WUTS_3 groups have not the same divergence than in the previous case. First, divergence is globally higher, thus indicating a more heterogeneous microscopic behavior. This newspaper should then be analyzed at a lower level of representation than *The Guardian*. Moreover, events are not reported in the same way, or with the same intensity, depending on the newspaper editorial policies.

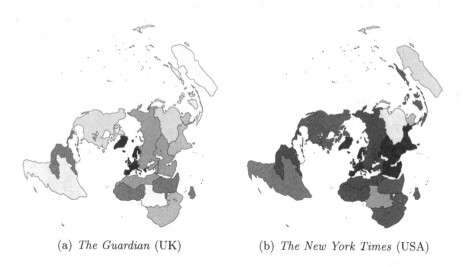

(a) *The Guardian* (UK)   (b) *The New York Times* (USA)

Fig. 6. Spatial variations of the KL divergence for groups of the WUTS_3 organization (the darker, the higher)

We do not aim at making explicit the various positive and negative factors explaining the citation number (*e.g.*, geographical, cultural, historical factors [15, 16]), but at showing that groups should be chosen with respect to the dataset. In our case, this is partly correlated with the source of the information. As a consequence, if an analyst uses distributed probes to observe a MAS, she does not want to use only one abstraction pattern to summarize the information.

This is consistent with the *subjectivist* account of emergence, according to which emergent phenomena strongly rely on the observation process [17].

4.3   Divergence of Groups Varies Over Time

Figure 7 presents the time variation of the KL divergence (thick red line) of the Northern America group (compounded of the USA, the Canada and the Mexico agents). Each value has been computed at the week level, by comparing the interpreted and the observed citation numbers for the countries of this group. The graph shows that a group with a globally low divergence through time can nonetheless be the source of significant information losses during specific time periods For example, the highest peak which appears in October 2012 is explained by the US presidential elections: during this time period, the USA agent accumulated much more citations than usually, whereas the Canada and the Mexico agents did not. Consequently, the use of the Northern America group to represent events led to a massive information loss. As a result, the choice of the representation level should also fit with the analyzed time period.

Moreover, the graph in Fig. 7 shows that the divergence variations are not strictly correlated with the variations of the analyzed variable (dashed blue line): an increasing of the observed-to-expected ratio of citation number does not implies an increasing of the divergence, and conversely. Hence, the citation number is not a sufficient criterion to evaluate the information content of organizations, by contrast with divergence.

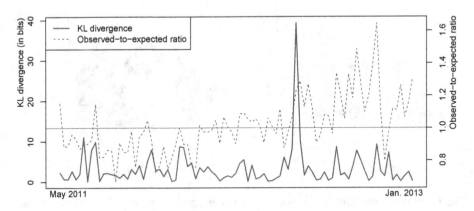

Fig. 7. Time variation of the KL Divergence and the observed-to-expected ratio of citation number of the Northern America group for *The Guardian* (Color figure online)

4.4   Divergence is Correlated with the Shape of Groups

The purpose of this third experiment is to compare two mesoscopic agent organizations: WUTS_2 and UNEP_reg (see Fig. 8). First, a global comparison indicates which organization minimizes the KL divergence. In order to compare the results from different newspapers, the information loss induced by organizations is normalized by the total citation number of the corresponding newspaper (in the following array, "b/c" stands for "bits/citation"):

	The Vancouver Sun	The Daily Mail	The Ph. Daily Inquirer
WUTS_2	1.80 b/c	1.46 b/c	2.07 b/c
UNEP_reg	1.57 b/c	1.51 b/c	2.26 b/c

It appears that, both for *The Daily Mail* and *The Philippine Daily Inquirer*, divergence is slightly lower for the WUTS_2 organization than for the UNEP_reg organization. Hence, if one should choose between these two, WUTS_2 should be preferred. However, for *The Vancouver Sun*, UNEP_reg is better. Once again, abstractions should then be chosen with respect to the source of information.

We can perform a more subtle analysis in order to determine the groups optimal shape. For example, we notice in Fig. 8 that U22 = W22 ∪ Mexico and W21 = U21 ∪ Mexico. Hence, one may ask "what is the best location of the Mexico agent?" Should it be aggregated with the Northern America group (W21/U21) or with the Latin America group (W22/U22)? For *The Daily Mail*, we have:

$$\mathrm{loss}(\mathtt{W21}) + \mathrm{loss}(\mathtt{W22}) = 0.048\ \mathrm{b/c} \quad < \quad 0.055\ \mathrm{b/c} = \mathrm{loss}(\mathtt{U21}) + \mathrm{loss}(\mathtt{U22})$$

Therefore, the observed-to-expected ratio of citation number of the Mexico agent is closer to those of the Northern America group. Mexico should be aggregated accordingly. This technique allows to evaluate the geographical abstractions used by the experts in terms of information content and to choose their optimal shape for the macroscopic analysis of a given dataset.

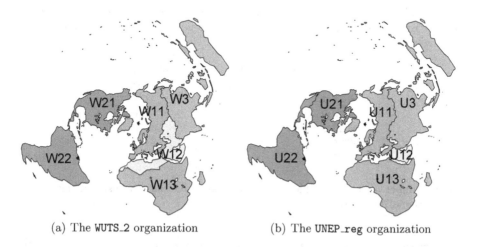

(a) The WUTS_2 organization   (b) The UNEP_reg organization

Fig. 8. Two organizations of the agent space in six similar (but not equivalent) groups: locations of the Northern Africa, the Western Asia and the Mexico subgroups differ

5 The Complexity Reduction Induced by the Aggregation

The information content is never increased by the aggregation process: for any pair of disjoint groups, we have: $\text{loss}(G_1 \cup G_2) \geq \text{loss}(G_1) + \text{loss}(G_2)$. Hence, if we only rely on KL divergence, the more detailed representation is always the best one. This is why we need a measure that also expresses what one *gains* by aggregating the microscopic data. To do so, this section presents two measures of *complexity reduction*. They estimate the information quantity that one saves by encoding a group G rather than its underlying agents:

$$\text{gain}(G) = \left(\sum_{a \in G} Q(a) \right) - Q(G)$$

where Q estimates the quantity of information needed to represent the agent a or the group G.

5.1 Number of Encoded Values

One way of measuring information quantities consists in estimating the number of bits needed to encode the values of a given representation. We may assume that it is constant for each agent or group: $Q(a) = Q(G) = q$, where q depends on the data type of the encoded values. Hence, for a group G, we have:

$$\text{gain}(G) = (|G| - 1) \times q$$

This function gives a basic complexity measure that fits well with classic visualization techniques (as for the maps in this paper) since the number of displayed values defines the granularity of the visualization. For example, according to the map expected complexity, the user can determine the number of groups that should be displayed. Figure 9 gives the organization size (number of groups) and the associated gain of each organizational level of the WUTS hierarchy.

Organization	$|O|$	$\text{gain}(O)/q$
WUTS_0	1	192
WUTS_1	3	190
WUTS_2	7	186
WUTS_3	17	176
WUTS_4	36	157
WUTS_5	193	0

Fig. 9. Number of encoded values and complexity reduction of the six organizational levels of the WUTS hierarchy

However, all groups do not contain the same number of agents. Thus, Fig. 10 gives, for each level of the WUTS hierarchy, the size (number of agents) of three disjoint high-level groups of countries: Euro-Africa, Americas and Asia-Pacific.

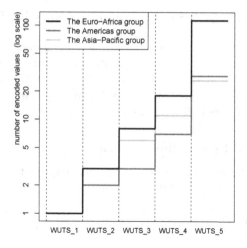

Fig. 10. Number of encoded values associated to the three groups of the WUTS_1 level

The user may want to adapt the representational level of these three groups depending on the amount of detail she expects for the corresponding geographical areas. The following section presents a criterion that automatically combines KL divergence and complexity reduction to adapt the size of groups depending on their quality, thus leading to multiresolution organizations.

5.2   Shannon Entropy

The number of encoded values only depends on the groups partitioning proposed by a given organization. In contrast, Shannon entropy also depends on the *variable distribution*. It is a classical complexity measure that is consistent with KL divergence (it can be defined as "the divergence from the uniform distribution"). Briefly, entropy evaluates the quantity of information needed to encode the countries associated to *each citation* (and not to encode the citation number associated to *each agent*). Based on Shannon's formula [3], we define the *entropy reduction* (or *gain*, in bits) of a group G as follows:

$$\text{gain}(G, t) = v(G, t) \, \log_2 v(G, t) \; - \; \sum_{a \in G} v(a, t) \, \log_2 v(a, t)$$

The choice of either one of these two complexity measures depends on the performed analysis. *Shannon entropy* should rather be used for the visualization of individuated citations, whereas *the number of encoded values* is more consistent with the visualization of aggregated values. In any case, the techniques presented in this paper are meant to be generic. They can be used with any complexity measure as long as it fits with some basic algebraic properties (see [18] for details).

6 Multiresolution Representation of Spatial Systems

As a conclusion to previous sections, finding a *good* organization relies on two aspects: complexity reduction (or *gain*), quantifying the granularity of the macroscopic representation, and KL divergence (or *loss*), quantifying the amount of information that has been lost during the aggregation process. Choosing an organization thus consists in finding a compromise between these two aspects.

6.1 Parametrized Information Criterion

We define a *parametrized Information Criterion* to express the trade-off between complexity reduction and information loss of a group G:

$$\mathrm{pIC}(G, t) \;=\; p \times \frac{\mathrm{gain}(G, t)}{\mathrm{gain}(A, t)} \;-\; (1 - p) \times \frac{\mathrm{loss}(G, t)}{\mathrm{loss}(A, t)}$$

where $p \in [0, 1]$ is a parameter used to balance the trade-off. For $p = 0$, maximizing the pIC is equivalent to minimizing the loss: the observer wants to be as precise as possible (microscopic level). For $p = 1$, she wants to be as simple as possible (full aggregation). When p varies from 0 to 1, a whole class of nested organizations arises. The observer has to choose the ones that fulfill her requirements, between the expected amount of details and the computational resources available for the analysis.

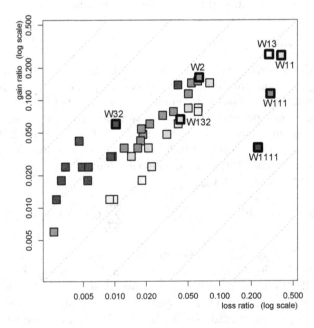

Fig. 11. Comparison of the ratio of information loss and the ratio of complexity reduction (logarithmic scales) for the groups of the WUTS hierarchy applied to the spatial data presented in Fig. 2 (Color figure online)

Figure 11 represents the groups of agents of the WUTS hierarchy (squares) depending on the two criteria that have been previously defined: the ratio of *KL divergence* $(\mathrm{loss}(G,t)/\mathrm{loss}(A,t))$ and the ratio of *encoded values reduction* $(\mathrm{gain}(G,t)/\mathrm{gain}(A,t))$. In this plot, quality groups are easily spotted:

- The closer to the *bottom right* corner the group is (red squares), the higher the information loss is relatively to the complexity reduction. This is for example the case of the Northern Europe group (W1111): the unexpected citation number of the Norway agent makes the group very heterogeneous (see events described in Subsect. 3.1). Higher-level European groups (W111 and W11) also induces a significant information loss (their are close to the right side). Thus, avoiding such groups during the aggregation ensures that we preserve the details regarding this significant microscopic variation.

- On the contrary, the closer to the *top left* corner the group is (green squares), the more the information loss is compensated by the complexity reduction. This is the case of the Americas group and the South Pacifica group (resp. W2 and W32). This indicates that these representation levels are particularly interesting to provide with a synthetic view of the system. These groups indeed correspond to homogeneous geographical areas where no significant event has occurred during the observed time period (see map in Fig. 2).

- The Horn of Africa group (W132) has a better gain/loss ratio than the higher-level Sub-Saharan Africa group (W13). This indicates that, if some details are necessary to analyze the events occurring in Africa, the Horn of Africa group can however be described as a whole, without giving more details regarding this particular area. Hence, by choosing groups depending on their gain/loss ratio, the observer can represent the system with several spatial granularities in order to perfectly fit with the microscopic data.

This method allows to spot interesting groups of agent to build a synthetic but consistent macroscopic representation of the system. The rest of this section proposes an algorithm to automatize this evaluation process and to find the combinations of groups (the organizations) that jointly optimize the two criteria.

6.2 Organizations Within a Hierarchy

Given a value of the trade-off parameter p, optimal organizations are those that maximize the parametrized information criterion. Clustering techniques using *gain* and *loss* measures as distances could find such optimal partitions. However, results may have very little meaning for the MAS analysis since agents would be aggregated regardless of their location within the system. In contrast, we assume that, in most spatial MAS, there is a correlation between topology and behavior. Hence, we propose that organizations should fit with the topological constraints defined by the domain experts. In other agent-based applications, such constraints may also be derived from *semantic* properties of the system (and not necessarily *topological* properties).

In this section, we consider hierarchically organized MAS. A *hierarchy H* is a set of nested groups, defined from the microscopic level (each agent is a group) to the whole MAS (only one group). The number of possible multiresolution organizations within such a hierarchy *exponentially* depends on the number of groups and the number of levels. For UNEP (196 groups arranged in 4 levels) and WUTS (231 groups arranged in 6 levels), we respectively have 1.3×10^6 and 3.8×10^{12} possible organizations. Finding the best one can thus be computationally expensive in case of large-scale systems. The algorithm below finds topologically-consistent organizations that maximize our parametrized information criterion. Its complexity *linearly* depends on the number of groups in the hierarchy (respectively 196 and 231 groups) by doing a depth-first search within the branches of the hierarchy. Indeed, according to the *sum property* [14] of the defined information-theoretic measures, each branch can be independently evaluated (see [19] for details).

Algorithm that linearly finds optimal organizations within a hierarchy

Require: A hierarchy H and a trade-off parameter p in $[0,1]$.
Ensure: An organization made of groups in H that maximizes the pIC.
1: **function** FINDOPTIMALORGANIZATION(H, p)
2:      **if** H contains only one group G **then return** $\{G\}$
3:      $G \leftarrow$ biggest group of H
4:      bestMicroOrganization $\leftarrow \emptyset$
5:      **for each** direct subhierarchy S of H **do**
6:          $aux \leftarrow$ FINDOPTIMALORGANIZATION(S, p)
7:          $bestMicroOrganization \leftarrow$ UNION($bestMicroOrganization, aux$)
8:      **end for**
9:      **if** pIC of $\{G\}$ > pIC of $bestMicroOrganization$ **then return** $\{G\}$
10:     **else return** $bestMicroOrganization$
11: **end function**

6.3 Hierarchical Aggregation to Build Spatial Macro-Representations

The above algorithm has been ran on the WUTS hierarchy for the articles published by *The Guardian* during July 2011 (see Fig. 2). The plot in Fig. 12 gives the complexity reduction and the information loss associated to the organizations provided by the algorithm depending on the trade-off parameter p specified by the observer. For $p = 0$, the optimal organization corresponds to the microscopic representation (no information loss and no complexity reduction). As p increases, groups of countries are chosen within the WUTS hierarchy in order to focus on significant events. The observer can adjust the granularity of the generated representation depending on the expected level of detail. Figures 13 and 14 present two organizations respectively preserving at least 50 % and 70 % of the microscopic information: $\mathrm{loss}(O,t)/\mathrm{loss}(A,t) < 0.5$ and $\mathrm{loss}(O,t)/\mathrm{loss}(A,t) < 0.3$ respectively for $p < 0.86$ and $p < 0.43$ (see Fig. 12).

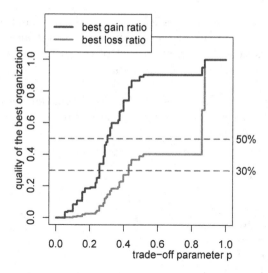

Fig. 12. Ratio of complexity reduction (*gain*) and information *loss* of the optimal organizations of the spatial data presented in Fig. 2 according to the trade-off parameter

The map in Fig. 13 represents the observed-to-expected ratio according to 17 groups of countries, two of which are high-level groups where the observed citation number is quite close to the expected value (the Americas group and the Asia-Pacific group). No significant event has been spotted at this level of detail. By contrast, two events are immediately highlighted in Europe and in Africa. They correspond to the most unexpected citation numbers for which an aggregation would induce a misleading information loss (at most 40 % of information loss when $p < 0.85$ and at least 68 % when $p > 0.85$, see Fig. 12).

1. The map in Fig. 13 displays the national details regarding agents of the Northern Europe group. Among them, the observed citation number of the Norway agent is 3.7 times higher than expected. The observer is thus informed that a significant event took place at the national level during July 2011 (the two terrorist attacks of the 22th, *cf.* Subsect. 3.1).

2. This map also displays some details regarding the Sub-Saharan Africa group, but only at a mesoscopic level constituted of 4 intermediary groups. Among them, the observer notices that the citation number of the Horn of Africa group is 1.8 times higher than expected. As the national details are not represented for this particular group, the observer may consider that this aggregated value is – at least at first glance – a good approximation of the underlying values. She concludes that the observed-to-expected ratio of citation number is *uniformly high* in the Horn of Africa. This group thus highlights an event that occurred in an extended geographical area (the food crisis that has been declared at the beginning of July 2011, *cf.* Subsect. 3.1).

The aggregation algorithm thus allows to highlight significant events that occur at different granularities of the system's spatial organization by building readable multiresolution representations out of the hierarchical structure. This map thus constitutes a reasonable "first approximation" to describe the news published by *The Guardian* during the month of July 2011. However, depending on the analyzed events and the expected level of detail, the observer may adapt the granularity of such a representation by adjusting the trade-off parameter (see the map presented in Fig. 14).

Fig. 13. Optimal geographical organization preserving at least 50 % of the microscopic information ($p < 0.86$)

The map below is a little bit more detailed than the previous one, in particular for countries of the `Asia-Pacific` group and those of the `Western Africa` group. Some other – less significant – microscopic events are thus represented:

3. The severe floods that occurred in Thailand at the end of July 2011[5].
4. The development cooperation project that began the 16th of July between the European Union and the Republic of Guinea-Bissau.

However, in this second map, the `Horn of Africa` group of agents is still aggregated. This is only when the observer asks for at least 83 % of the microscopic information ($p < 0.39$) that the algorithm provides the national details regarding this geographical area. In that way, the algorithm can adapt the generated representations to the observer's requirements.

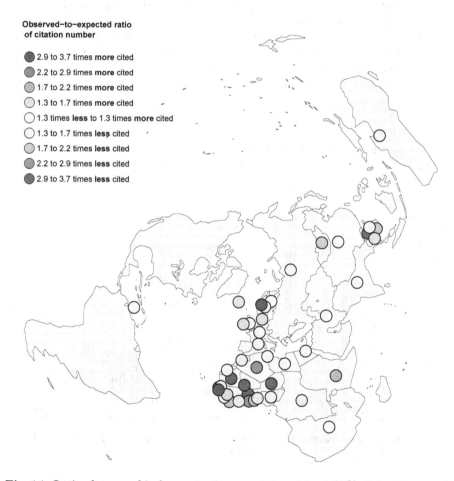

Fig. 14. Optimal geographical organization preserving at least 70 % of the microscopic information ($p < 0.43$)

---

[5] http://en.wikipedia.org/wiki/2011_Thailand_floods

7   Generalization to Temporal Aggregation

The time series in Fig. 15 provides with the week-level variations of the observed-to-expected ratio of citation number of the `Greece` agent by *The Guardian* between the 3rd of May 2011 and the 20th of January 2013. Peaks of unexpected citation number reveal significant events in Greece recent history. In the following, we will focus on three particular events:

1. The peak of citation that appears at the beginning of November 2011 is explained by the announcement the 31st of October of a referendum regarding the setting up of an austerity plan to reduce the Greek public debt. This announcement is widely reported by the media. On the 4th of November, the Minister of Finance announces the referendum withdrawal and the Prime Minister George Papandreou arranges, on the same evening, a vote of confidence in the Parliament that could lead to his resignation.
2. The peak that appears in the middle of May 2012 is explained by the failure of the legislative elections that are held the 6th of May and concluded the 16th by the establishment of an interim government until the organization of new elections.
3. The peak that appears at the end of June 2012 is explained by the holding of these second legislative elections on the 17th of June[6].

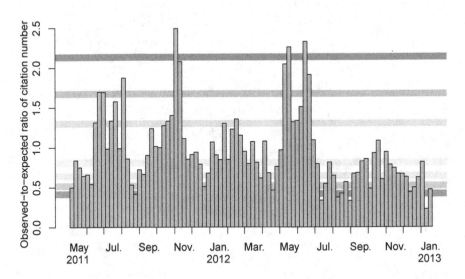

Fig. 15. Microscopic time series (week level) of the observed-to-expected ratio of citation number of the `Greece` agent by *The Guardian*

---

[6] See the Wikipedia page dedicated to the Greek government-debt crisis to get more details regarding the chronology of these political events:
http://en.wikipedia.org/wiki/Greek_government-debt_crisis

In this section, the aggregation technique is applied to the system's temporal dimension. In this case, macroscopic events refer to time periods during which the citation number of a given country (or group of countries) has been much higher than expected. Such periods – that breaks with the system's stable state – may also be defined at different time scales (days, weeks, month, years, and so on). When interpreting an aggregated time period, the observer can use – as for the spatial aggregation – the marginal values to distribute the aggregated citation number over the underlying microscopic time periods (*i.e.*, the week level) depending on the total number of citations on these time periods. For a microscopic time period t in an aggregated time period $T' \subset T$, the interpreted citation number is:

$$Q(a,t) = v(a,T') \frac{v(A,t)}{v(A,T')}$$

This interpreted value is then compared to the observed value: $P(a,t) = v(a,t)$, according to KL divergence:

$$\text{loss}(a,T') = \sum_{t \in T'} P(a,t) \log_2 \left(\frac{P(a,t)}{Q(a,t)} \right)$$

$$= \sum_{t \in T'} v(a,t) \log_2 \left(\frac{v(a,t)\, v(A,T')}{v(a,T')\, v(A,t)} \right)$$

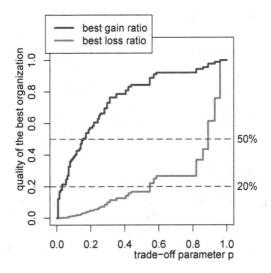

Fig. 16. Ratio of complexity reduction (*gain*) and information *loss* of the optimal partitions of the temporal data presented in Fig. 15 according to the trade-off parameter

The optimization of the corresponding *parametrized Information Criterion* (see Subsect. 6.1) can be achieved by the algorithm of Jackson *et al.* [20]. It consists in slicing the time series in intervals that maximize a given fitness function.

We have shown in [19] that this time-aggregation algorithm is part of a larger class of algorithms – including the space-aggregation algorithm proposed in this paper – that consist in computing the optimal partitions of a dataset under some constraints (*e.g.*, hierarchical organization for spatial aggregation, ordered dataset for temporal aggregation). The time-aggregation algorithm of Jackson *et al.* is thus exploited as previously to build multiresolution representations of the agent dynamics.

The plot in Fig. 16 gives the complexity reduction and the information loss of the time partitions provided by the algorithm depending on the trade-off parameter p. Series in Figs. 17 and 18 present two such optimal partitions respectively preserving at least 80 % and 50 % of the week-level information (resp. $p < 0.55$ and $p < 0.89$). The time series in Fig. 17 thus summarizes the microscopic data by aggregating groups of weeks for which the observed-to-expected ratio of citation number is quite homogeneous. Even if the result contains less details, it still provides significant information for the analysis of Greek news. In particular, the three significant aforementioned peaks are highlighted and easily spotted by the observer. Moreover, the variations between the peaks are synthetically represented. This allows to describe the system dynamics according to macroscopic time periods.

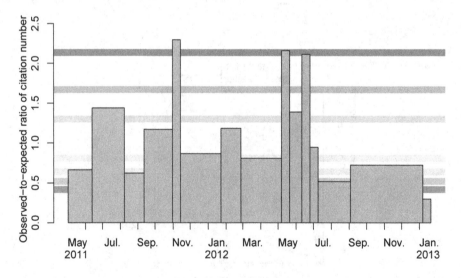

Fig. 17. Optimal temporal partition preserving at least 50 % of the microscopic information ($p < 0.55$)

The time series in Fig. 18 provides with an even more aggregated representation of time variations. Only the most significant events are represented:

1. The first peak, corresponding to the announcement the 31st of October of a Greek referendum, is strongly highlighted by the aggregation process.

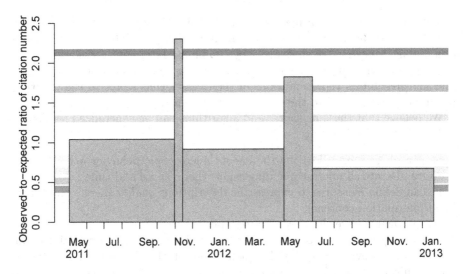

Fig. 18. Optimal temporal partition preserving at least 80 % of the microscopic information ($p < 0.89$)

2. The two peaks of May and June 2012, respectively corresponding to the legislative elections of the 6th of May and the 17th of June, are aggregated together, now constituting a unique homogeneous time period of 7 weeks. The corresponding event is interpreted as "the Greek legislative elections of 2012", without giving the week-level details of this month-scale event. The aggregation process thus provides consistent temporal abstractions to describe and analyze Greek events at a higher level of representation.
3. This time series also gives an interesting macroscopic information: the citation number of the `Greece` agent has been globally decreasing during the observation period. This could be explained by the declining media interest regarding the Greek crisis after the arrival in news of other economical crises concerning European countries such as Spain and Italy. The aggregation process thus allows to represent the system's temporal dynamics at several time-scales by highlighting significant macroscopic variations. If we had the sufficient temporal depth, we would be able to identify a year-long time period in Greek history corresponding to "the government-dept crisis" described as a whole.

8 Conclusion and Perspectives

The design and debugging of complex MAS need abstraction tools to work at a higher level of representation. However, such tools have to be developed and exploited with the greatest precaution in order to preserve useful information regarding the system behavior and to guarantee that generated representations are not misleading for the observer. To that extent, this paper focuses on aggregation techniques for large-scale MAS and gives clues to estimate their quality in term of complexity and information content. They are applied to the geographical and temporal aggregation of international relations through the point of view of on-line newspapers. We show that, by combining information-theoretic measures, one can give interesting feedback regarding the use of abstractions and build multiresolution representations of the dataset that adapt to the effective information content and to the observer's requirements.

We believe that these measures and algorithms can be generalized to a large class of MAS, provided that:

- one can observe and describe the agents microscopic behavior according to several discretized microscopic dimensions (here: space and time);
- one can define measures to express the descriptions *quality* (here: complexity and information content);
- these measures have the *sum property* [14];
- the semantic and topological properties of the aggregated dimensions can be used to provide meaningful abstractions for the domain experts (here: hierarchical organizations and order of time).

Future work will apply these techniques to other dimensions of the analysis: *e.g.* for aggregation of newspapers, thematic aggregation, multidimensional aggregation [19]. Besides this work, we are currently exploiting these techniques for performance visualization of large-scale distributed systems [21]. This kind of application shows that our techniques can be scaled up to one million agents.

Acknowledgement. This work was partially funded by the ANR CORPUS GEOMEDIA project (ANR-GUI-AAP-04). We would like to thank Claude Grasland, Timothée Giraud and Marta Severo for their work on this project; and Lucas M. Schnorr for his close participation to previous work.

References

1. Elmqvist, N., Fekete, J.: Hierarchical aggregation for information visualization: overview, techniques, and design guidelines. IEEE Trans. Visual Comput. Graphics **16**(3), 439–454 (2010)
2. Kullback, S., Leibler, R.: On information and sufficiency. Ann. Math. Stat. **22**(1), 79–86 (1951)
3. Shannon, C.: A mathematical theory of communication. Bell Syst. Tech. J. **27**(379–423), 623–656 (1948)
4. Railsback, S.F., Lytinen, S.L., Jackson, S.K.: Agent-based simulation platforms: review and development recommendations. Simulation **82**, 609–623 (2006)
5. Van Liedekerke, M.H., Avouris, N.M.: Debugging multi-agent systems. Inf. Softw. Technol. **37**, 103–112 (1995)
6. Búrdalo, L., Terrasa, A., Julián, V., García-Fornes, A.: A tracing system architecture for self-adaptive multiagent systems. In: Demazeau, Y., Dignum, F., Corchado, J.M., Pérez, J.B. (eds.) Advances in PAAMS. AISC, vol. 70, pp. 205–210. Springer, Heidelberg (2010)

7. Tonn, J., Kaiser, S.: ASGARD – a graphical monitoring tool for distributed agent infrastructures. In: Demazeau, Y., Dignum, F., Corchado, J.M., Pérez, J.B. (eds.) Advances in PAAMS. AISC, vol. 70, pp. 163–173. Springer, Heidelberg (2010)

8. Sharpanskykh, A., Treur, J.: Group abstraction for large-scale agent-based social diffusion models with unaffected agents. In: Kinny, D., Hsu, J.Y., Governatori, G., Ghose, A.K. (eds.) PRIMA 2011. LNCS, vol. 7047, pp. 129–142. Springer, Heidelberg (2011)

9. Iravani, P.: Multi-level network analysis of multi-agent systems. In: Iocchi, L., Matsubara, H., Weitzenfeld, A., Zhou, C. (eds.) RoboCup 2008. LNCS (LNAI), vol. 5399, pp. 495–506. Springer, Heidelberg (2009)

10. Peng, W., Grushin, A., Manikonda, V., Krueger, W., Carlos, P., Santos, M.: Graph-based methods for the analysis of large-scale multiagent systems. In: AAMAS'09, IFAAMAS, pp. 545–552 (2009)

11. Gil-Quijano, J., Louail, T., Hutzler, G.: From biological to Urban cells: lessons from three multilevel agent-based models. In: Desai, N., Liu, A., Winikoff, M. (eds.) PRIMA 2010. LNCS, vol. 7057, pp. 620–635. Springer, Heidelberg (2012)

12. Grasland, C., Didelon, C.: Europe in the World - Final Report. Volume 1, ESPON Project 3.4.1 (2007)

13. United Nations Environment Programme: Global Environmental Outlook: environment for development. Volume 4, Nairobi (2007)

14. Csiszár, I.: Axiomatic characterizations of information measures. Entropy 10(3), 261–273 (2008)

15. Galtung, J., Ruge, M.H.: The structure of foreign news: the presentation of the Congo, Cuba and Cyprus crises in four Norwegian newspapers. J. Peace Res. 2(1), 64–91 (1965)

16. Koopmans, R., Vliegenthart, R.: Media attention as the outcome of a diffusion process–a theoretical framework and cross-national evidence on earthquake coverage. Eur. Sociol. Rev. 27(5), 636–653 (2011)

17. Deguet, J., Demazeau, Y., Magnin, L.: Element about the emergence issue: a survey of emergence definitions. ComPlexUs 3, 24–31 (2006)

18. Lamarche-Perrin, R., Vincent, J.M., Demazeau, Y.: Informational measures of aggregation for complex systems analysis. Technical report RR-LIG-026, Laboratoire d'Informatique de Grenoble, France (2012)

19. Lamarche-Perrin, R., Demazeau, Y., Vincent, J.M.: The best-partitions problem: how to build meaningful aggregations. In: Proceedings of the 2013 IEEE/WIC/ACM International Conference on Intelligent Agent Technology (IAT'13), Atlanta, GA, USA, pp. 399–404. IEEE Computer Society (2013)

20. Jackson, B., Scargle, J.D., Barnes, D., Arabhi, S., Alt, A., Gioumousis, P., Gwin, E., Sangtrakulcharoen, P., et al.: An algorithm for optimal partitioning of data on an interval. IEEE Signal Process. Lett. 12(2), 105–108 (2005)

21. Lamarche-Perrin, R., Schnorr, L.M., Vincent, J.M., Demazeau, Y.: Evaluating trace aggregation for performance visualization of large distributed systems. In: Proceedings of the 2014 IEEE International Symposium on Performance Analysis of Systems and Software (ISPASS'14), Monterey, CA, USA (2014)

Understanding the Role of Emotions in Group Dynamics in Emergency Situations

Alexei Sharpanskykh[1(✉)] and Kashif Zia[2]

[1] Faculty of Aerospace Engineering, Delft University of Technology,
Kluyverweg 1, 2629 HS Delft, The Netherlands
o.a.sharpanskykh@tudelft.nl
[2] COMSATS Institute of Information Technology, Abbottabad, Pakistan
kashifzia@ciit.net.pk

Abstract. Decision making under stressful circumstances, e.g., during evacuation, often involves strong emotions and emotional contagion from others. In this paper the role of emotions in social decision making in large technically assisted crowds is investigated. For this a formal, computational model is proposed, which integrates existing neurological and cognitive theories of affective decision making. Based on this model several variants of a large scale crowd evacuation scenario were simulated. By analysis of the simulation results it was established that (1) human agents supported by personal assistant devices are recognised as leaders in groups emerging in evacuation; (2) spread of emotions in a crowd increases the resistance of agent groups to opinion changes; (3) spread of emotions in a group increases its cohesiveness; (4) emotional influences in and between groups are, however, attenuated by personal assistant devices, when their number is large.

Keywords: Crowd evacuation · Cognitive modelling · Ambient intelligence · Multi-agent simulation

1 Introduction

Decision making under stressful circumstances, e.g., during evacuation, often involves strong emotions and emotional contagion from others [1, 6]. More generally, it is widely recognised in cognitive and neurological literature that emotions influence human decision making [2, 9, 12]. However, quantifying this influence is a challenging task. Previously, human decision making has been considered as entirely rational and has been modelled using economic utility-based theories [19, 20]. Purely rational decision making models were disapproved by many empirical studies (see e.g., [20]). However, devising a better alternative addressing the limitations of these models by combining cognitive (rational) and affective (emotional) aspects still remains a big challenge.

To address this challenge several computational models were proposed [10, 27, 29], which use variants of the OCC model developed by Ortony, Clore and Collins [23] as a basis. The OCC model postulates that emotions are valenced reactions to events, agents, and objects, where valuations are based on similarities between achieved states and goal states. Thus, emotions in this model have a cognitive origin. In contrast to

R. Kowalczyk et al. (Eds.): TCCI XV, LNCS 8670, pp. 28–48, 2014.
DOI: 10.1007/978-3-662-44750-5_2

these approaches, we employ a neurological fundament, on which a model of social decision making is built. This model exploits some of the principles underlying the OCC model but embeds them in a neurological context. By taking a neurological perspective and incorporating cognitive and affective elements in one integrated model, a more realistic and deeper understanding of the internal processing underlying human decision making in social situations can been achieved. This gives a richer type of model than models purely at the cognitive level, or diffusion (contagion) models at the social level abstracting from internal processing, for example, as addressed in [17]. More specifically, options in decision making involving sequences of actions are modelled using the neurological theory of simulated behaviour (and perception) chains proposed by Hesslow [16]. Moreover, the emergence of emotional states in these behavioural chains is modelled using emotion generation principles described by Damasio [7–9]. Evaluation of sensory consequences of actions in behavioural chains, also uses elements borrowed from the OCC model. Different types of emotions can be distinguished and their roles in the decision making clarified. Two types of emotions – hope and fear – are particularly relevant in the context of crowd evacuation. The emergence and dynamics of these two emotions are addressed in the model presented in the paper.

Evaluation of decision options and the emotions involved in it usually have a strong impact from the human's earlier experiences. In the proposed model for social decision making, this form of adaptivity to past experiences is also incorporated based on neurological principles. In such a way elements from neurological, affective and cognitive theories were integrated in the adaptive agent model proposed.

Usually decision making occurs in a social context (e.g., a group of people). People influence others and are influenced by others. In many studies on emotional decision making the social context is either ignored [27, 29] or comprises a small group of individuals [17]. In this paper we investigate social decision making in large crowds of people. The effects of emotional decision making on a large scale (a crowd) may differ significantly from the ones on a small scale (an individual or a small group).

Due to the ubiquitous use of personal communication devices (e.g., mobile phones), which often play a prominent role in emergency situations, also such devices need to be included in the model as information sources. Both researchers and authorities envision an important contribution of such and more intelligent assistant devices to monitoring and control of large mass events [13]. Thus, in the model some of the human agents are equipped with technical devices called personal assistants, able to receive information relevant for decision options from other devices.

In the literature [1, 26, 28, 30] it is indicated that people often form spontaneous groups during evacuation. On the one hand, dynamic formation of groups is recognised as a prerequisite for efficient evacuation [1, 30]. On the other hand, large uncontrolled groups may cause clogging of paths and increase panic [1, 26]. In this paper, *a group* is defined by a set of human agents, supporting the same decision option and located closely to each other in the physical space. To investigate the role of emotions in the formation and dynamics of groups, 5 hypotheses were formulated, which are discussed in the following.

In [24] the possession of knowledge is identified as a strong power basis in social groups, especially when they are situated in environments with scarce and uncertain information. In line with this argument, the following hypothesis is formulated:

Hypothesis 1: Human agents equipped with personal assistants, who obtain up-to-date information about the environment, are recognised as leaders in groups emerging in evacuation.

The next hypothesis is a known observation from the social psychology literature confirmed by empirical studies (see e.g., [22]):

Hypothesis 2: Emotions increase the consistency of social decision making and the robustness of a group against external perturbations (e.g., receipt of inconsistent information from strangers).

The third hypothesis follows from the second one.

Hypothesis 3: Emotions arising in social decision making increase the group cohesiveness.

Hypothesis 4: The higher the penetration rate of personal assistants, the less the influence of emotions on the group dynamics.

The last hypothesis is related to *the large group effect* known for social emergency systems [1]:

Hypothesis 5: Evacuation with larger groups proceeds more slowly (less efficiently) than with smaller groups.

The hypotheses were tested by agent-based simulation based on the proposed emotional decision making model in the context of a large scale crowd evacuation scenario. To validate the hypotheses the two-sample t-test was applied [32]. By analysis of the simulation results all the hypotheses were confirmed.

The paper is organised as follows. A case study is introduced in Sect. 2. The general modelling principles on which the proposed model is based are described in Sect. 3. A detailed formalisation of the proposed model for the evacuation scenario is provided in Sect. 4. The simulation and verification results for the hypotheses are presented in Sect. 5. Finally, Sect. 6 concludes the paper.

2 Case Study

In the simulation study we focussed on evacuation of a train station. To ensure that the simulation setting is a true representative of reality, a real CAD design of an existing Austrian main railway station was incorporated to generate the space along with observed population statistics.

The station in the simulation model had 3 exits with different flow capacities. Exit E13 has largest capacity equal to a width of 7 cells followed by Exit E15 consisting of width equal to 5 cells. Exit SC1 has least width equal to 2 cells. The station was populated randomly with 1000 agents representing humans, from which a number of agents depending of the simulation trial (1 %, 5 % or 10 %) were equipped with personal assistants (see Fig. 1). In Fig. 1, three different colours representing agents heading towards three exits respectively (blue towards Exit E15, green towards Exit E13 and yellow towards Exit SC1) are shown. Out of a total population of 1000 agents, 1 % (with red labels) are equipped with personal assistants.

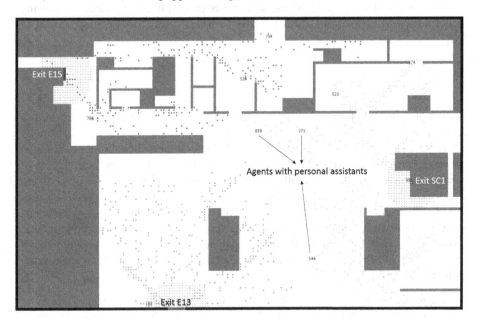

Fig. 1. A train station represented in the simulation environment with coloured dots representing agents heading towards three exits (Color figure online)

All personal assistants constantly received information about the degree of clogging of each exit from a global 'evacuation control unit'. This information was assumed to be measured by a technology mounted on each exit. Furthermore, it is assumed that the global control unit provides reliable, up-to-date information to all personal assistants without any noise.

Each personal assistant had a location map used to transform the coordinates of an exit to the desired orientation to move. Thus, agents with personal assistants had direct access to information essential for successful evacuation, which they could propagate further by interaction with other agents.

Agents can interact with each other *non-verbally* by spreading emotions and intentions to choose particular exits, and *verbally* by communicating information about the states of the exits. The agents without devices were free to decide whether or not to follow agents with personal assistants or to rely on their own beliefs and exit choices. It is important to stress that the grouping effect is not encoded in our model explicitly, but emerges as a result of complex decision making by agents.

To verify the hypotheses formulated in the introduction, three variants of the scenario were introduced, which were simulated:

Variant 1: Agents generate and exchange both information and emotions during the social decision making.

Variant 2: Agents generate both emotions and information, but exchange only information.

Variant 3: Agents generate and exchange only information.

The simulation of all variants of the scenario is based on a social decision making model described in Sect. 4, which relies on a neurological fundament described in Sect. 3.

3 Theoretical Basis

Considering options and evaluating them is viewed as a central process in human decision making. An option is a sequence of actions to achieve a goal, as in planning. To model considering such sequences, from the neurological literature the *simulation hypothesis* proposed by Hesslow [16] was adopted. Based on this hypothesis, chains of behaviour can be simulated as follows: some situation elicits activation of s1 in the sensory cortex that leads to preparation for action r1. Then, associations are used such that r1 will generate s2, which is the most connected sensory consequence of the action for which r1 was generated. This sensory state serves as a stimulus for a new response, and so on. In such a way long chains of simulated responses and perceptions representing plans of action can be formed. These chains are simulated by an agent internally as follows:

'An anticipation mechanism will enable an organism to simulate the behavioural chain by performing covert responses and the perceptual activity elicited by the anticipation mechanism. Even if no overt movements and no sensory consequences occur, a large part of what goes on inside the organism will resemble the events arising during actual interaction with the environment.' [16]

As reported in [16], behavioural experiments have demonstrated a number of striking similarities between simulated and actual behaviour.

Hesslow argues in [16] that the simulated sensory states elicit emotions, which can guide future behaviour either by reinforcing or punishing simulated actions. However, specific mechanisms for emotion elicitation are not provided. This gap can be filled by combining the simulation hypothesis with a second source of knowledge from the neurological area: Damasio's emotion generation principles based on *(as-if) body loops*, and the principle of *somatic marking* [2, 8]. These principles were adopted to model evaluation of options.

Damasio [7–9] argues that sensory or other representation states of a person often induce emotions felt within this person, according to a *body loop* described by the following causal chain:

sensory state $\rightarrow$ preparation for the induced bodily response $\rightarrow$ induced bodily response $\rightarrow$ sensing the bodily response $\rightarrow$ sensory representation of the bodily response $\rightarrow$ induced feeling

As a variation, an *as if body loop* uses a direct causal relation as a shortcut in the causal chain: preparation for the induced bodily response $\rightarrow$ sensory representation of the induced bodily response. The body loop (or 'as if body loop') is extended to a recursive body loop (or recursive 'as if body loop') by assuming that the preparation of the bodily response is also affected by the state of feeling the emotion as an additional causal relation: feeling → preparation for the bodily response. Thus, agent emotions are modelled based on reciprocal causation relations between emotion felt and body states. Following these emotion generation principles, an 'as if body' loop can be incorporated in a simulated behavioural chain as shown in Fig. 2 (left). Note that based on the sensory states different types of emotions may be generated.

In the *OCC model* [23] a number of cognitive structures for different types of emotions are described. By evaluating sensory consequences of actions s1, s2, ..., sn from Fig. 2 using cognitive structures from the OCC model, different types of emotions can be distinguished. More specifically, the emergence of hope and fear in agent decision making in an emergency scenario will be considered in Sect. 4. The OCC model has been extensively used for representing emotions in diverse ambience intelligence frameworks. For example, in [33], using the OCC model emotions are generated that influence decision making of and negotiation between agents in a group. No neurological or psychological validity of the model is asserted in this work. Moreover, the knowledge about emotional influences on social processes in ambient intelligence environments is still rather limited. To the best of our knowledge, influence of emotions on such aspects as group cohesiveness and robustness of social decision making in an ambient intelligence setting has not been studied before.

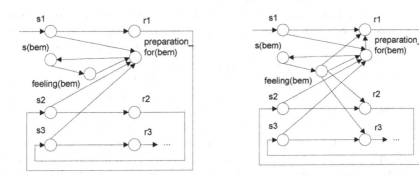

Fig. 2. Simulation of a behavioural chain extended with an 'as if body' loop with emotional state bem (left) and with emotional influences on preparation states (right)

Hesslow argues in [16] that emotions may reinforce or punish simulated actions, which may transfer to overt actions, or serve as discriminative stimuli. Again, specific mechanisms are not provided. To fill this gap the Damasio's *Somatic Marker Hypothesis* was adopted. This hypothesis provides a central role in decision making to emotions felt. Within a given context, each represented decision option induces (via an emotional response) a feeling which is used to mark the option. For example, a strongly negative somatic marker linked to a particular option occurs as a strongly negative feeling for that option. Similarly, a positive somatic marker occurs as a positive feeling for that option. Damasio describes the use of somatic markers in the following way:

'the somatic marker (..) forces attention on the negative outcome to which a given action may lead, and functions as an automated alarm signal which says: beware of danger ahead if you choose the option which leads to this outcome. The signal may lead you to reject, *immediately*, the negative course of action and thus make you choose among other alternatives. (…) When a positive somatic marker is juxtaposed instead, it becomes a beacon of incentive.' [9, pp. 173–174]

To realise the somatic marker hypothesis in behavioural chains, emotional influences on the preparation state for an action are defined as shown in Fig. 2 (right). Through these connections emotions influence the agent's readiness to choose the option. From a neurological perspective, the impact of a sensory state to an action preparation state via the connection between them in a behavioural chain will depend on how the consequences of the action are felt emotionally.

As neurons involved in these states and in the associated 'as if body' loop will often be activated simultaneously, such a connection from the sensory state to the preparation to action state may be strengthened based on a general *Hebbian learning* principle [14, 15] that was adopted as well. It describes how connections between neurons that are activated simultaneously are strengthened, similar to what has been proposed for the emergence of mirror neurons; e.g., [18, 25]. Roughly spoken this principle states that connections between neurons that are activated simultaneously are strengthened. From a Hebbian perspective, strengthening of connections as mentioned in case of positive valuation may be reasonable, as due to feedback cycles in the model structure, neurons involved will be activated simultaneously.

Thus, by these processes an agent differentiates options to act based on the strength of the connection between the sensory state of an option and the corresponding preparation to an action state, influenced by its emotions. The option with the highest activation of preparation is chosen to be performed by the agent.

As also used as an inspiration in [17], in a social context, the idea of somatic marking can be combined with recent neurological findings on the *mirroring function* of certain neurons (e.g., [18, 25]). Mirror neurons are neurons which, in the context of the neural circuits in which they are embedded, show both a function to prepare for certain actions or bodily changes and a function to mirror similar states of other persons. They are active not only when a person intends to perform a specific action or body change, but also when the person observes somebody else intending or performing this action or body change. This includes expressing emotions in body states, such as facial expressions. The mirroring function relates to decision making in two different ways. In the first place *mirroring of emotions* indicates how emotions felt in different individuals about a certain considered decision option mutually affect each other, and, assuming a context of somatic marking, in this way affect how by individuals decision options are valuated in relation to how they feel about them. A second way in which a mirroring function relates to decision making is by applying it to the *mirroring of intentions or action tendencies* of individuals (i.e., preparation states for an action) for the respective decision options. This may work when by verbal and/or nonverbal behaviour individuals show in how far they tend to choose for a certain option. In the computational model introduced below in Sect. 4 both of these (emotion and preparation) mirroring effects are incorporated.

4 Modelling Emotional Decision Making

First, in Sect. 4.1 a modelling language is described used for formalisation of the model. Then, the formal model is provided in Sect. 4.2.

4.1 The Modelling Language

To specify dynamic properties of a system, the order-sorted predicate logic-based language called LEADSTO is used [4]. This language satisfies essential demands for dynamic modelling of agent systems in natural domains. In particular, it allows the possibility of both discrete and continuous modelling of a system at different aggregation levels. Furthermore, it has numerical expressivity for modelling systems with explicitly defined quantitative relations presented by difference or differential equations. Moreover, for specifying qualitative aspects of a system, LEADSTO is able to express logical relationships between parts of a system.

Dynamics in LEADSTO is represented as evolution of states over time. A state is characterized by a set of properties that do or do not hold at a certain point in time. To specify state properties for system components, ontologies are used which are defined by a number of sorts, sorted constants, variables, functions and predicates (i.e., a signature). For every system component A a number of ontologies can be distinguished: the ontologies IntOnt(A), InOnt(A), OutOnt(A), and ExtOnt(A) are used to express respectively internal, input, output and external state properties of the component A. Input ontologies contain elements for describing perceptions of an agent from the external world, whereas output ontologies describe actions and communications of agents. For a given ontology Ont, the propositional language signature consisting of all state ground atoms based on Ont is denoted by APROP(Ont). State properties are specified based on such ontology by propositions that can be made (using conjunction, negation, disjunction, implication) from the ground atoms. Then, a *state* S is an indication of which atomic state properties are true and which are false: S: APROP (Ont) → {true, false}.

LEADSTO enables modeling of direct temporal dependencies between two state properties in successive states, also called *dynamic properties*. A specification of dynamic properties in LEADSTO is executable and can be depicted graphically. The format is defined as follows. Let $\alpha1$ and $\alpha2$ be state properties of the form 'conjunction of atoms or negations of atoms', and e, f, g, h non-negative real numbers. In the LEADSTO language the notation $\alpha1 \twoheadrightarrow_{e, f, g, h} \alpha2$ means: if state property $\alpha1$ holds for a certain time interval with duration g, then after some delay (between e and f) state property $\alpha2$ will hold for a certain time interval of length h (Fig. 3). When e = f = 0 and g = h = 1, called standard time parameters, we shall write $\alpha1 \twoheadrightarrow \alpha2$. To indicate the type of a state property in a LEADSTO property we shall use prefixes input(c), internal(c) and output(c), where c is the name of a component. Consider an example dynamic property:

input(A)|observation_result(fire) ⇸ output(A)| performed(runs_away_from_fire)

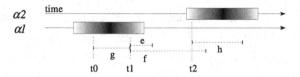

Fig. 3. Timing relationships for LEADSTO expressions.

Informally, this example expresses that if agent A observes fire during some time unit, then after that A will run away from the fire during the following time unit.

In addition, LEADSTO allows expressing mathematical operations, e.g., has_value $(x, v) ⇸_{e, f, g, h}$ has_value(x, v*0.25).

Next, a *trace* or *trajectory* γ over a state ontology Ont is a time-indexed sequence of states over Ont (where the time frame is formalised by the real numbers). A LEADSTO expression α1 $⇸_{e, f, g, h}$ α2, holds for a trace γ if:

$$∀t1 \, [∀t[t1 - g ≤ t < t1 ⇒ α1 \text{ holds in } γ \text{ at time } t]$$
$$⇒ ∃d[e ≤ d ≤ f \, \& \, ∀t' \, [t1 + d ≤ t' < t1 + d + h ⇒ α2 \text{ holds in } γ \text{ at time } t']]$$

To specify the fact that a certain event (i.e., a state property) holds at every state (time point) within a certain time interval a predicate holds_during_interval(event, t1, t2) is introduced. Here event is some state property, t1 is the beginning of the interval and t2 is the end of the interval.

An important use of the LEADSTO language is as a specification language for simulation models. As indicated above, on the one hand LEADSTO expressions can be considered as logical expressions with a declarative, temporal semantics, showing what it means that they hold in a given trace. More details on the semantics of the LEADSTO language can be found in [4].

4.2   The Computational Model

Depending on a situational context an agent determines a set of applicable options to satisfy its goal. In the case study the goal of each agent is to get outside of the train station in the fast possible way. This is achieved by an agent by moving towards the exit that provides for fastest evacuation as it perceived by the agent. Evacuation options are represented internally in agents by one-step simulated behavioural chains, based on the neurological theory by Hesslow [16] (see Fig. 4). Note that if more than one exit is known to the agent, then in each option representation the preparation state corresponding to the option's exit is generated. Computationally, alternative options considered by an agent are being generated and evaluated in parallel.

According to the Somatic Marker Hypothesis [8], each represented decision option induces (via an emotional response) a feeling(s) which is used to mark the option. For example, a strongly positive somatic marker linked to a particular option occurs as a strongly positive feeling for that option. The decision options from our study evoke two types of emotions - fear and hope, which are often considered in the emergency context. To realise the somatic marker hypothesis in behavioural chains, emotional influences on the preparation state for an action are defined as shown in Fig. 4. Through these connections emotions influence the agent's readiness to choose the option.

In Fig. 4 the burning station situation elicits activation of the state srs(evacuation_required) in the agent's sensory cortex that leads to preparation for action preparation_for(move_to(E)). Here E is one of the exits of the station. Furthermore, this preparation state is affected by the sensory representations of the perceived preparation of the neighbouring agents for the action and of the emotions felt towards the option.

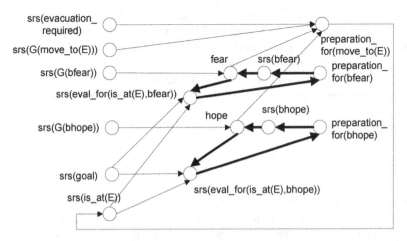

Fig. 4. The emotional decision making model for the option to move to exit E.

Formally:

srs(evacuation_required, V1) & srs(fear, V2) & srs(hope, V5) & srs(G(move_to(E)), V3) & preparation_for(move_to(E), V4)

$\rightarrow$ preparation_for(move_to(E), V4 + γ(h(V1, V2, V3, V5) – V4)Δt),

where G(move_to(E) is the aggregated preparation of the neighbouring agents to action move_to(E), h(V1, V2, V3, V5) is a combination function:

h(V1, V2, V3, V5) = β (1 – (1 – V1)V2(1 – V3)(1 – V5)) + (1 – β) V1 V3 V5(1 – V2)

Here β is a parameter that reflects the agent's predisposition to think positively (β > 0.5) or negatively (β < 0.5). Parameter γ reflects the agent's rate of change of its state.

The option with the highest activation of preparation is chosen to be performed by the agent.

Then, associations are used such that preparation_for(move_to(E)) will generate srs (is_at(E)), which is the most connected sensory consequence of the action move_to(E). The strength of the link between a preparation for an action and a sensory representation of the effect of the action (see Fig. 4) is used to represent the confidence value of the agent's belief that the action leads to the effect. This is modelled by the following formal property:

preparation_for(move_to(E), V) & connection_between_strength(preparation_for(move_to(E)), srs (is_at(E)), ω) ⇸ srs(is_at(E), ωV)

The simulated sensory states elicit emotions, which guide agent behaviour either by reinforcing or punishing simulated actions. By evaluating sensory consequences of actions in simulated behavioural chains using cognitive structures from the OCC model [23], different types of emotions can be distinguished. As a simulated behavioural chain is a kind of a behavioural projection, cognitive structures of prospect-based emotions (e.g., fear, hope, satisfaction, disappointment) from [23] are particularly relevant for the evaluation process. In our study two types of emotions - fear and hope – are distinguished. According to [23], the intensity of fear induced by an event depends on the degree to which the event is undesirable and on the likelihood of the event. The intensity of hope induced by an event depends on the degree to which the event is desirable and on the likelihood of the event. Thus, both emotions are generated based on the evaluation of a distance between the effect states for the action from an option and the agent's goal state.

In particular, the evaluation function for hope in the evacuation scenario is specified as

$$eval(g, \ is\_at(E)) = \omega,$$

where ω is the confidence value for the belief about the accessibility of exit E, which is an aggregate of the agent's estimation of the distance to the exit and the degree of clogging of the exit. Although it is assumed that the distances to the exits are known to the agents, the information about the degree of clogging of the exits is known only to technology-equipped agents.

Emotions emerge and develop in dynamics of reciprocal relations between cognitive and body states of a human [7, 8]. These relations, omitted in the OCC model, are modelled from a neurological perspective using Damasio's principles of 'as-if body' loops and somatic marking described in Sect. 3. The 'as-if body' loops for hope and fear emotions are depicted in Fig. 4 by thick solid arrows. These loops are formalised by the properties provided below.

The evaluation properties for fear and for hope of the effect of action move_to(E) compared with the goal state goal is specified formally as:

srs(goal, V1) & srs(is_at(E), V2) & srs(fear, V3) &
connection_between_strength(preparation_for(move_to(E)), srs(is_at(E)), V4) &
srs(eval_for(is_at(E), bfear), V5)
⇸ srs(eval_for(is_at(E), bfear), V5 + γ(h(V4*f(goal, is_at(E)), V3) – V5) Δt),

where f(goal,is_at(E)) = |V1-V6|, V6 = eval(goal, is_at(E)), and
h(V1, V2) = β (1-(1-V1)(1-V2)) + (1-β) V1 V2.

srs(goal, V1) & srs(is_at(E), V2) & srs(hope, V3) &
connection_between_strength(preparation_for(move_to(E)), srs(is_at(E)), V4) &
srs(eval_for(is_at(E), bhope), V5)
$\twoheadrightarrow$ srs(eval_for(is_at(E), bhope), V5 + γ(h(V4* f(goal, is_at(E)), V3) – V5) Δt),

where f(goal, is_at(E)) = 1-|V1-V6|, and V6 = eval(goal, is_at(E)).

The evaluation of the effects of the action for a particular emotional response to an option determines the intensity of the emotional response:

srs(eval_for(is_at(E), bhope), V1) $\twoheadrightarrow$ preparation_for(bhope, V1)

srs(eval_for(is_at(E), bfear), V1) $\twoheadrightarrow$ preparation_for(bfear, V1)

The agent perceives its own emotional response and creates the sensory representation state for it:

preparation_for(bhope, V) $\twoheadrightarrow$ srs(bhope, V)

preparation_for(bfear, V) $\twoheadrightarrow$ srs(bfear, V)

Finally the dynamics of the emotional states are formalised as follows:

srs(bhope, V1) & srs(G(bhope), V2) & srs(hope, V3) $\twoheadrightarrow$ srs(hope, V3 + γ(h(V1, V2) – V3) Δt)),

where h(V1,V2) is a combination function defined above.

srs(bfear, V1) & srs(G(bfear), V2) & srs(fear, V3) $\twoheadrightarrow$ srs(fear, V3 + γ(h(V1, V2) – V3) Δt)),

The social influence on the individual decision making is modelled based on the mirroring function [18] of preparation neurons in humans. It is assumed that the preparation states of an agent for the actions and for emotional responses for the options are body states that can be observed with a certain intensity or strength by other agents from the neighbourhood. Furthermore, it is assumed that an agent is able to observe preparation states of other agents in its neighbourhood specified by radius r. Note that the agent's neighbourhood changes while the agent moves.

The *contagion strength* of the interaction from agent A_2 to agent A_1 for a preparation state p is defined as follows:

$$\gamma_{pA_2A_1} = \varepsilon_{pA_2} \cdot trust(A_1,A_2) \cdot \alpha_{pA_2A_1} \cdot \delta_{pA_1}$$

Here ε_{pA2} is the personal characteristic expressiveness of the sender (agent A_2) for p, δ_{pA1} is the personal characteristic openness of the receiver (agent A_1) for p.

Trust is an attitude of an agent towards an information source that determines the extent to which information received by the agent from the source influences agent's belief(s). The trust to a source builds up over time based on the agent's experience with the source. In particular, when the agent has a positive (negative) experience with the source, the agent's trust to the source increases (decreases). Currently experiences are restricted to information experiences only. An information experience with a source is evaluated by comparing the information provided by the source with the agent's beliefs about the content of the information provided. The experience is evaluated as positive (negative), when the information provided by the source is confirmed by (disagree with) the agent's beliefs. The following property describes the update of trust of agent A_i to agent A_j based on information communicated by A_j to A_i about the degree of clogging of exit E:

trust(A_i, A_j, V1) & communicated_from_to(A_j, A_i, clogging(E, V2)) & belief(A_i, clogging(E, V3))
$\rightarrow$ trust(A_i, A_j, V1 + γ_{tr}*(V3/(1 + e$^\alpha$) − V1)),

here $\alpha = -\omega1*(1-|V2-V3|)$, parameter $\omega1$ determines the steepness of change of the trust state.

An agent A_i perceives the joint attitude of the crowd towards each option by aggregating the input from all agents in its neighbourhood $\aleph$:
(a) the aggregated neighbourhood's preparation to each action p is expressed by the following dynamic property:

$\wedge_{Aj\in\aleph}$ internal(A_j)|preparation_for(p, V_j) $\rightarrow$ internal(A_i)|srs(G(p), Σ $_{j\neq i}$ γ_{pAjAi} V_j /Σ $_{j\neq i}$ $\gamma_{pAjAi}\varepsilon_{pAj}$)

(b) the aggregated neighbourhood's preparation to the emotional responses (hope and fear) for each option:

$\wedge_{Aj\in\aleph}$internal(A_j)|preparation_for(bhope,V_j)$\rightarrow$internal(A_i)|srs(G(bhope),$\Sigma_{j\neq i}$ γ_{beAjAi} V_j/Σ $_{j\neq i}$ $\gamma_{beAjAi}\varepsilon_{beAj}$)

$\wedge_{Aj\in\aleph}$internal(A_j)|preparation_for(bhope,V_j)$\rightarrow$internal(A_i)|srs(G(bhope),$\Sigma_{j\neq i}$ γ_{beAjAi} V_j/Σ $_{j\neq i}$ $\gamma_{beAjAi}\varepsilon_{beAj}$)

The Hebbian learning principle for links connecting the sensory representation of options with preparation states for subsequent actions in the simulation of a behavioural chain is formalised as follows (cf. [14, 15]):

connection_between_strength(srs(evacuation_required), preparation_for(move_to(E)), V1) & srs (srs(evacuation_required), V2) & preparation_for(move_to(E), V3)
$\rightarrow$ connection_between_strength(srs(evacuation_required), preparation_for(move_to(E)), V1 + (η V2 V3 (1 − V1) − ξV1)Δt),

where η is a learning rate and ξ is an extinction rate.

5   Simulation Results

The model was implemented in the Netlogo simulation tool [31] by cellular automata. In this tool the environment is represented by a set of connected cells, where moveable agents (turtles) reside. Cells can be walkable (open space and exits) and not-walkable (concrete, partitions, walls). Each cell of the environment is accessible from all the exits.

The three variants of the model described in Sect. 2 were implemented as 3 simulation conditions:

Condition 1: Agents generate and exchange both information and emotions during the social decision making.
Condition 2: Agents generate both emotions and information, but exchange only information.
Condition 3: Agents generate and exchange only information.

Since the model contains stochastic elements, 50 trials were performed for each simulation setting with 1000 heterogeneous agents with the parameters drawn from the ranges of uniformly distributed values as indicated in Table 1 below to represent a diversity of agent types that may occur in real emergency situations. It is assumed that the agents have average to high expressiveness and openness. The agents do not have a strong predisposition to think positively or negatively (β) in the simulation. The agents have an average to high rate of change of their states (γ). The agents have an average learning rate (η) and a low extinction rate (ξ), as often assumed in neurological models. It is assumed that humans trust technology in the same manner as to human strangers. A human agent has a low initial trust value to all other agents; it increases (decreases) slowly ($\omega 1 = 9$) its trust to an agent after a positive (negative) experience with the agent.

Table 1. Ranges and values of the agent parameters used in the simulation.

ε for all states from all agents	δ for all states from all agents	β	γ	η	ξ	Δt	r	$\omega 1$	Initial trust to all agents
[0.7, 1]	[0.7, 1]	[0.55, 0.7]	[0.7, 1]	0.6	0.1	1	10	9	[0.1, 0.3]

In the following simulation results and testing of the hypotheses formulated in Sect. 1 are discussed. To test the hypotheses, the simulation traces generated for each condition were analysed using the TTL Checker Tool [5].

To evaluate *Hypothesis 1* two evaluation metrics were introduced: *following index (fi)*, which reflects the degree of following of technology-assisted agents by other agents, and *group size (gs)*. As shown below, the metrics are defined per a technology-assisted agent L (i.e., fi_L, gs_L) and by taking the mean over all technology-assisted agents (i.e., fi, gs). The following index is defined as follows:

$$fi_L = 1/|N| \cdot \sum_{A \in N} |F_{A,L}|/(t\_last - t\_first_A), \qquad fi = \sum_{i \in LEAD} fi_i/|LEAD|,$$

where t_first$_A$ is such that

∃o1:INFO at(communicated_from_to(L, A, inform, o1), t_first$_A$) & ∀t:TIME, o:INFO t < t_first$_A$ & ¬at(communicated_from_to(L, A, inform, o), t);
N = {a | t_first$_A$ is defined}; $F_{A,L}$ = {t | t ≥ t_first$_A$ & ∃d1,d2: DECISION at(has_preference_for(A, d1), t) & at(has_preference_for(L, d2), t) & d1 = d2 & at(distance_between(A, L) < dist_threshold, t) }, t_last is the time point when L is evacuated, LEAD is the set of all technology-assisted agents.

The group size is defined as follows:

$$gs_L = \sum_{t=1..t\_last} FT_{L,t}/t\_last, \qquad gs = \sum_{i \in LEAD} gs_L/|LEAD|,$$

where $FT_{L,t}$ = {ag | t ≥ t_first$_{ag}$ & ∃d1,d2: DECISION at(has_preference_for(ag, d1), t) & at (has_preference_for(L, d2), t) & d1 = d2 & at(distance_between(A, L) < dist_threshold, t) }.

The obtained results are summarised in Table 2. As one can see from the table, the emergence of groups with agents equipped with personal assistants as guiding leaders occurs in all conditions (fi > 0), thus, the hypothesis 1 is confirmed.

In *Condition 1* the most clogged exit throughout the simulation is Exit SC1, as it is the closest exit to most of the agents (Fig. 5a). As information about clogging of other exits spreads through the population of agents, the clogging of Exit SC1 decreases, but still remains higher than the clogging of other exits. Agents react to the change of clogging of the exits by changing their preferred exits (Fig. 5b). The amount of agents aiming at exit SC1 decreases throughout the simulation, whereas the numbers of agents choosing E15 and E13 fluctuate depending on the situation around these exits.

Table 2. The simulation results for 50 simulation trials for three simulation conditions. Standard deviation is provided in brackets.

Coefficient	Condition 1	Condition 2	Condition 3
fi	0.42 (0.15)	0.33 (0.11)	0.21 (0.11)
gs	27 (8.1)	15 (5.5)	11(3.2)
si$_{exit1}$	0.12 (0.03)	0.32 (0.04)	0.65 (0.07)
si$_{exit2}$	0.12 (0.04)	0.23 (0.05)	0.45 (0.08)
si$_{exit3}$	0.13 (0.04)	0.21 (0.07)	0.29 (0.07)
ci	1.5 (0.4)	1.9 (0.7)	7.1 (0.7)

Information about the exits received by the agents influences their emotional states (Fig. 6). The technology-assisted agents, who receive information about exits first, change their emotions more rapidly than the agents without such devices (cf. the dynamics of hope in Fig. 6a and b). Furthermore, information provided by the technology-assisted agents spreads rapidly and is readily accepted by other agents, as can be seen from the similarity of the dynamics of the emotions in Fig. 6a and b.

To verify *Hypothesis 2* a smoothness degree of the preparation for each action (i.e., move to exit E) averaged over all agents is determined in each simulation trial (*smoothness index* (si$_E$)):

$$si_E = \sum_{t=1...t\_last-1,\ a\in Np_t,E,a}/|N|,$$

$$\text{with } p_{t,E,a} = \begin{cases} |v_{t+1,E,a}-v_{t,E,a}|, & \text{when } |v_{t+1,E,a}-v_{t,E,a}| \geq \varepsilon \\ 0, & \text{when } |v_{t+1,E,a}-v_{t,E,a}| < \varepsilon \end{cases}$$

Here N is the set of all agents, $v_{t,E,a}$ is the value of preparation_for(move_to(E)) for agent a at time point t; ε is a threshold for distinguishing small changes from large changes; ε is taken 0.1 for the analysis.

Thus, the smoothness index depends on the rate of change of the agent's opinion based on incoming information. This index indicates the robustness of a group of agents to messages provided by agents outside the group, which support a decision

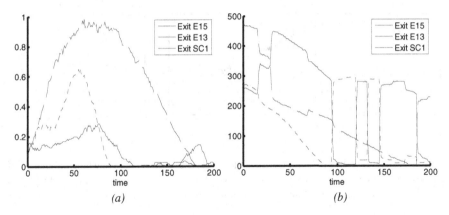

Fig. 5. (a) The change of the degree of clogging of each exit over time in *Condition 1*; (b) The change of numbers of agents heading to each exit in *Condition 1*.

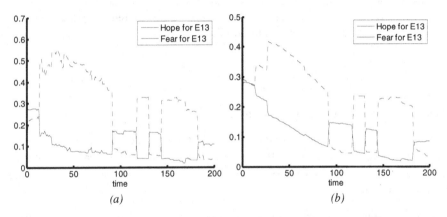

Fig. 6. The emotional response toward the option to follow exit E13 averaged over technology-assisted agents (a) and over the agents without devices (b).

option different from the one currently supported by the group. Note that a group is defined by a set of human agents, supporting the same decision option and located closely to each other in the physical space. In the evacuation scenario this occurs when the situation around an exit(s) changes. Then, the agents with personal assistants receive new information, based on which they may change their decisions. Further, these agents spread new information to other agents in their neighbourhood. If besides information also emotions are being spread (see Table 2, condition 1 and Fig. 7a), the population of agents change their decisions gradually. When emotions are generated, but are not being spread, the group becomes less robust to changes and reacts more abruptly to incoming messages (see Table 2, condition 2 and Fig. 7b).

In the situation when emotions are not generated, the agents in a group change their decisions frequently, rapidly and drastically (see Table 2, condition 2 and Fig. 7b). Such a form of behaviour is highly unrealistic for human beings.

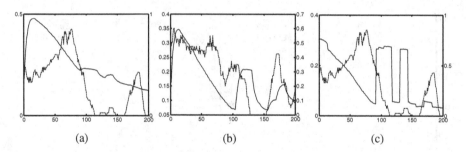

Fig. 7. The change of the preparation to move to exit E15 averaged over the whole population of agents (solid line; left vertical axis), and the change of the degree of clogging of exit E15 (dotted line; right vertical axis) in condition 1(a), condition 2(b) and condition 3(c); the horizontal line is time.

Thus, the outcomes of the simulation support *Hypothesis 2* that generation and spread of emotions increase the consistency of social decision making.

To verify *Hypothesis 3* the metrics called *change index* (ci), reflecting the frequency of group change by an agent, was introduced.

It is defined by:

$$ci_L = 1/|N| \sum_{A \in N} |S_{A,L}|, \qquad\qquad ci = \sum_{i \in LEAD} ci_i/|LEAD|,$$

where LEAD is the set of all agents with personal assistants,

$S_{A,L}$ = {t | (t∈ $F_{A,L}$ & (t + 1) ∉ $F_{A,L}$) OR ((t + 1)∈ $F_{A,L}$ & t ∉ $F_{A,L}$) }, and

$F_{A,L}$ = {t | t ≥ t_first$_A$ & ∃d1,d2:DECISION at(has_preference_for(A, d1), t) & at(has_preference_for (L, d2), t) & d1 = d2 & at(distance_between(A, L) < dist_threshold, t)}, at(X,t) denotes that X holds at time point t, and

t_first$_A$ is such that ∃o1:INFO at(communicated_from_to(L, A, inform, o1), t_first$_A$) & ∀t:TIME, o: INFO t < t_first$_A$ & ¬at(communicated_from_to(L, A, inform, o), t), and

N = {a | t_first$_A$ is defined}.

The average change index in *Condition 3* was 4.7 and 3.7 times higher than in *Conditions 1* and *2* respectively (Table 2, ci row). Thus, when emotions are not generated, agents are significantly less attached to their group than in the case when emotions are generated and being spread. The two-sample t-test performed on the outcomes of *Condition 3* and *Condition 1* and on the outcomes of *Condition 3* and *Condition 2* confirms *Hypothesis 3* with 95 % confidence.

To test *Hypothesis 4*, *Conditions 1* and *2*, with and without spread of emotions correspondingly, with the penetration rates of personal assistant devices equal to 1, 5 and 10 % were simulated 50 times each. Then, for each simulation case the mean values of the coefficients si_{exit1}, si_{exit2}, si_{exit3}, ci, fi, describing the dynamics of emerging groups, were determined. After that, the differences between the corresponding coefficients for *Conditions 1* and *2* were calculated averaged over 50 simulations (Table 3), which can be seen as measures of similarity of the group dynamics between the conditions.

Table 3. The differences between the group dynamics coefficients for *Conditions 1* and *2* for different penetration rates averaged over 50 simulations

Penetration rate, %	1	5	10
$<si_{exit1}^{cond2} - si_{exit1}^{cond1}>$	0.35	0.2	0.05
$<si_{exit2}^{cond2} - si_{exit2}^{cond1}>$	0.25	0.11	0.03
$si_{exit3}^{cond2} - si_{exit3}^{cond1}>$	0.21	0.08	0.03
$<ci^{cond2} - ci^{cond1}>$	0.9	0.4	0.1
$<fi^{cond2} - fi^{cond1}>$	0.12	0.09	0.04

The results in Table 3 indicate that with an increase of the number of personal assistant devices, the differences between *Conditions 1* and *2* become smaller. This can be explained as follows: Personal assistant devices support the consistency of social decision making by providing uniform information to human agents. When the number of the personal assistant devices becomes high, most of the human agents will be situated within the reach of such devices. In this case, the devices will (partially) overtake the role of emotions by providing information to human agents, which will increase the cohesiveness of groups and the consistency of their decision making. Because of this, the role of emotional influences, and thus differences between the *Conditions 1* and *2*, will be diminished. This supports *Hypothesis 4*.

To test *Hypothesis 5*, *Condition 1* was simulated 50 times with two more propagation radii: $r = 5$ and $r = 20$. It can be observed in Table 4 that the mean group size and the overall evacuation time grow with the increase of the interaction range. The two-sample t-test performed on the outcomes of two pairs of conditions - with the interaction range 5 and 10, and with the interaction range 10 and 20 - confirms *Hypothesis 5* with 95 % confidence.

Table 4. The mean overall evacuation time and the mean size of the groups emerging in the simulation of *Condition 1* with different interaction ranges. Standard deviation is provided in brackets.

Interaction range, r	5	10	20
Mean group size, gs	17 (4.8)	27 (8.1)	54 (10.2)
Mean overall evacuation time in seconds	156.2 (24.3)	164.1 (31.6)	201.4 (32.2)

6 Conclusion

Many empirical studies indicated [7, 9, 19, 22] that emotions play an important role in social decision making. In this paper the role of emotions in group dynamics in large crowds has been investigated. To this end, based on the literature from social psychology and domain knowledge five hypotheses were formulated. To verify these hypotheses a computational model for social decision making was developed. This model is based on a number of neurological theories and principles supplementing each other in a consistent manner. By simulation based on this model and performing the two-sample t tests on the results all these hypotheses were confirmed. In particular, human agents equipped with personal assistants were recognised as leaders in groups emerging in evacuation. Evacuation with larger groups proceeded more slowly than with smaller groups. Spread of emotions in a crowd increased resistance of agent groups to opinion changes. Acceptance of a different decision option occurred gradually, as also described in the literature [21, 22]. Furthermore, spread of emotions in a group increased its cohesiveness. This result is also supported by the literature (e.g., see [22]). Emotional influences were, however, attenuated by an increasing number of personal assistant devices.

The modelling perspective followed aims at a cognitive and affective modelling level, but takes inspiration from the underlying mechanisms as described at a neurological level. Modeling causal relations discussed in neurological literature in a cognitive/affective level model does not take specific neurons into consideration but uses more abstract mental states. This is a way to use results from the large and more and more growing amount of neurological literature for the cognitive/affective modelling level. This method can be considered as lifting neurological knowledge to a higher level of description. In a more detailed manner, in [3], such a perspective is discussed: '… we can expect that injection of some neurobiological details back into folk psychology would fruitfully enrich the latter, and thus allow development of a more fine-grained folk-psychological account that better matches the detailed functional profiles that neurobiology assigns to its representational states.' [3]. Here Bickle suggests that by relating a (folk) psychological to a neurobiological account, the psychological account can be enriched. The type of higher level model that results from adopting principles from the neurological level may inherit some characteristics from the neurological level. In particular this holds for the Hebbian learning principle adopted here. Another, even more basic element inherited from this 'lifting' perspective is the use of numbers to indicate the strength of the considered states. This is more common in neural modelling perspectives, but here also applied at a higher level. Such a gradual way of modelling allows for the type of cyclic and adaptive processes addressed here, which would be impossible using an approach based on a binary states.

To generate emotions the OCC model has been used in the paper. However, there also exist other approaches to emotional modelling, such as the basic emotions approach [34] and the dimensional approach [35]. The former approach is similar to the OCC model in distinguishing a set of basic emotions (e.g., happiness, anger). The latter approach distinguishes a few dimensions (e.g., valence and arousal) to characterise different emotions; e.g., fear is characterised by a negative valence and a high arousal. Both these approaches can be incorporated in our model by defining appropriate evaluation functions, as discussed in Sect. 4.2.

In the literature [11] it is recognized that humans often employ diverse emotion regulation mechanisms (e.g., to cope with fear and stress). These mechanisms involve interplay between cognitive and affective processes. In the future the proposed model will be extended with an emotion regulation component.

Furthermore, in real evacuation communication lines might be broken and information relay may be significantly delayed. Such scenarios were not considered in this paper and are left for future work.

Acknowledgement. One of the authors was supported by the Dutch Technology Foundation STW, which is the applied science division of NWO, and the Technology Program of the Ministry of Economic Affairs.

References

1. Barton, A.H.: Communities in Disaster: A Sociological Analysis of Collective Stress Situations. Doubleday, Garden City (1969)
2. Bechara, A., Damasio, A.: The somatic marker hypothesis: a neural theory of economic decision. Games Econ. Behav. **52**, 336–372 (2004)
3. Bickle, J.: Psychoneural Reduction: The New Wave. MIT Press, Cambridge (1998)
4. Bosse, T., Jonker, C.M., van der Meij, L., Treur, J.: A language and environment for analysis of dynamics by simulation. Int. J. AI Tools **16**, 435–464 (2007)
5. Bosse, T., Jonker, C.M., van der Meij, L., Sharpanskykh, A., Treur, J.: Specification and verification of dynamics in agent models. Int. J. Cooper. Inf. Syst. **18**(1), 167–193 (2009)
6. Bosse, T., Hoogendoorn, M., Klein, M.C.A., Treur, J., van der Wal, C.N., van Wissen, A.: Modelling collective decision making in groups and crowds: integrating social contagion and interacting emotions, beliefs and intentions. Auton. Agents Multi-Agent Syst. J. **27**, 52–84 (2013)
7. Damasio, A.: The Feeling of What Happens. Body and Emotion in the Making of Consciousness. Harcourt Brace, New York (1999)
8. Damasio, A.: The somatic marker hypothesis and the possible functions of the prefrontal cortex. Philos. Trans. R. Soc. Biol. Sci. **351**, 1413–1420 (1996)
9. Damasio, A.: Descartes' Error: Emotion, Reason and the Human Brain. Papermac, London (1994)
10. Baillie, P., Lukose, D.: An affective decision making agent architecture using emotion appraisals. In: Ishizuka, M., Sattar, A. (eds.) PRICAI 2002. LNCS (LNAI), vol. 2417, pp. 581–590. Springer, Heidelberg (2002)
11. Delgado, M.R., Phelps, E.A., Robbins, T.W.: Decision Making, Affect, and Learning: Attention and Performance XXIII. Oxford University Press, New York (2011)
12. Eich, E., Kihlstrom, J.F., Bower, G.H., Forgas, J.P., Niedenthal, P.M.: Cognition and Emotion. Oxford University Press, New York (2000)
13. Ferscha, A., Farrahi, K., van den Hoven, J., Hales, D., Nowak, A., Lukowicz, P., Helbing, D.: Socio-inspired ICT - Towards a socially grounded society-ICT symbiosis. Eur. Phys. J. Spec. Top. **214**, 401–434 (2012)
14. Gerstner, W., Kistler, W.M.: Mathematical formulations of Hebbian learning. Biol. Cybern. **87**, 404–415 (2002)
15. Hebb, D.O.: The Organisation of Behavior. Wiley, New York (1949)
16. Hesslow, G.: Conscious thought as simulation of behaviour and perception. Trends Cogn. Sci. **6**, 242–247 (2002)
17. Hoogendoorn, M., Treur, J., van der Wal, C., van Wissen, A.: Modelling the emergence of group decisions based on mirroring and somatic marking. In: Yao, Y., Sun, R., Poggio, T., Liu, J., Zhong, N., Huang, J. (eds.) BI 2010. LNCS (LNAI), vol. 6334, pp. 29–41. Springer, Heidelberg (2010)

18. Iacoboni, M.: Understanding others: imitation, language, empathy. In: Hurley, S., Chater, N. (eds.) Perspectives on Imitation: From Cognitive Neuroscience to Social Science, vol. 1, pp. 77–100. MIT Press, Perspectives on Imitation: From Cognitive Neuroscience to Social Science, vol. 1, pp. 77–100. MIT Press (2005)

19. Janis, I., Mann, L.: Decision Making: A Psychological Analysis of Conflict, Choice, and Commitment. The Free Press, New York (1977)

20. Kahneman, D., Slovic, P., Tversky, A.: Judgement Under Uncertainty - Heuristics and Biases. Cambridge University Press, Cambridge (1981)

21. Lewin, K.: Group Decision and Social Change. Holt, Rinehart and Winston, New York (1958)

22. Magee, J.C., Tiedens, L.Z.: Emotional ties that bind: the roles of valence and consistency of group emotion in inferences of cohesiveness and common fate. Pers. Soc. Psychol. Bull. **32**, 1703–1715 (2006)

23. Ortony, A., Clore, G.L., Collins, A.: The Cognitive Structure of Emotions. Cambridge University Press, New York (1988)

24. Raven, B.H.: The bases of power: origins and recent developments. J. Soc. Issues **12**(49), 227–251 (1992)

25. Rizzolatti, G., Craighero, L.: The mirror-neuron system. Annu. Rev. Neurosci. **27**, 69–92 (2004)

26. Saloma, C., Perez, G.J.: Herding in real escape panic. In: Proceedings of the 3rd International Conference on Pedestrian and Evacuation Dynamics. Springer, Heidelberg (2006)

27. Santos, R., Marreiros, G., Ramos, C., Neves, J., Bulas-Cruz, J.: Multi-agent approach for ubiquitous group decision support involving emotions. In: Ma, J., Jin, H., Yang, L.T., Tsai, J.J.-P. (eds.) UIC 2006. LNCS, vol. 4159, pp. 1174–1185. Springer, Heidelberg (2006)

28. Sharma, S.: Avatarsim. A multi-agent system for emergency evacuation simulation. J. Comput. Methods Sci. Eng. **9**(1), 13–22 (2009)

29. Steunebrink, B.R., Dastani, M., Meyer, J.-J.C.: A logic of emotions for intelligent agents. In: Proceedings of the 22nd Conference on Artificial Intelligence (AAAI 2007). AAAI Press, Menlo Park (2007)

30. Svenson, O., Maule, A.J. (eds.): Time Pressure and Stress in Human Judgment and Decision-Making. Plenum, New York (1993)

31. NetLogo tool. http://ccl.northwestern.edu/netlogo. Last accessed Nov 2010

32. Moore, D.S., McCabe, G.P.: Introduction to the Practice of Statistics. W. H. Freeman and Company, New York (2007)

33. Marreiros, G., Santos, R., Ramos, C., Neves, J., Novais, P., Machado, J., Bulas-Cruz, J.: Ambient intelligence in emotion based ubiquitous decision making. In: Augusto, J.C., Shapiro, D. (eds.) Proceedings of the 2nd Workshop on Artificial Intelligence Techniques for Ambient Intelligence (AITAmI), pp. 86–91 (2007)

34. Oatley, K., Johnson-Laird, P.N.: Towards a cognitive theory of emotions. Cogn. Emotion **1**, 29–50 (1987)

35. Vastfjall, D., Friman, M., Garling, T., Kleiner, M.: The measurement of core affect: a Swedish self-report measure. Scand. J. Psychol. **43**, 19–31 (2002)

Representation of the Agent Environment for Traffic Behavioral Simulation

Feirouz Ksontini[1,2]([envelope]), Stéphane Espié[2], Zahia Guessoum[3],
and René Mandiau[1]

[1] Université de Valenciennes et du Hainaut Cambrésis - LAMIH CNRS 8201,
59313 Valenciennes, France
{feirouz.ksontini,rene.mandiau}@univ-valenciennes.fr
[2] Université Paris-Est/IFSTTAR/IM, 14-20 bd Newton,
77447 Champs-sur-Marne, France
stephane.espie@iffstar.fr
[3] Université de Paris 6 - LIP6, 4 Place Jussieu, 75252 Paris, France
zahia.guessoum@lip6.fr

Abstract. The aim of this paper is to improve the validity of traffic simulations in (sub-)urban context, with a better consideration of driver behavior in terms of anticipation of positioning on the lanes and occupation of space. We introduce a model based on a multi-agent approach and the emergence concept. This model considers that each driver perceives the situation in an ego-centered way and readapts the road space using the virtual lane concept. We implement the model with the traffic simulation tool ArchiSim. The so obtained simulator intends to reproduce the observed behavior such as filtering between vehicles (two-wheels, emergency vehicles), repositioning on lanes when approaching the road intersections and "exceptional" situations (stranded vehicle or improperly parked, etc.).

Keywords: Multi-agent simulation · Road traffic simulation · Ego-centered environment representation · Virtual lanes

1 Introduction

Two kinds of approaches are proposed to simulate traffic and to study the related phenomena: mathematical and behavioral approaches. Commonly used traffic simulation tools are based on mathematical models which use different statistical laws resulting from measurements on the field. The so obtained laws rely on the physical characteristics of the road on which are made the measurements (*e.g.*, length, number of lanes, capacity). The limit of these models, based on aggregation of individual situations makes difficult the reproduction of the anticipation phenomena. The behavioral approach offers a solution when the aim of the simulation is to produce behaviors, realistic at both individual level and collective level. Traffic phenomena (*e.g.*, lanes occupancy, congestion) result from individual practices (*e.g.*, heterogeneous driver behaviors), interactions and the travel

© Springer-Verlag Berlin Heidelberg 2014
R. Kowalczyk et al. (Eds.): TCCI XV, LNCS 8670, pp. 49–68, 2014.
DOI: 10.1007/978-3-662-44750-5_3

context (geometry and structure of the road, regulation, etc.). In this context, multi-agent simulation allows to simulate traffic system actors by autonomous agents in more realistically way, thanks to the decision-making process.

We use a multi-agent behavioral approach to describe the road traffic simulation. This approach was developed over the past twenty years by IFSTTAR[1] in the traffic simulation tool *ArchiSim* [1,2]. The aim is to produce the observed practices, and in particular, to simulate the behavior related to the anticipation phenomena (*e.g.*, anticipation of positioning on the lanes) as well as the occupation of road space, particularly in context of high traffic density in urban areas. Our model focuses on situations such as the filtering maneuvers between vehicles (two-wheels), the readapting road space in approaching and in the intersections, the specific events (stranded vehicle or improperly parked, etc.), the dynamic lane allocation, etc.

Existing simulation models do not consider all the above mentioned phenomena [3–7]. So, the related simulations do not always reproduce the real observed phenomena. Previous work have proposed solutions for the particular case of two-wheels [6,7]. Our purpose is to develop a generic model for the above mentioned practices taking into account the specificities of each driver for the most varied possible situations. We so present a method to build a generic and ego-centered environment representation which uses the concept of virtual lanes and relies on the results of some driving psychological studies [8–12].

The paper is organized as follows. Section 2 deals with the multi-agent simulation studies related to our problematic. Section 3 focuses on the phenomena of readapting the road space and the issue of ego-centered environment representation. Section 4 describes briefly the *ArchiSim*architecture. In Sect. 5, we present our ego-centered representation model of the environment. Section 6 is devoted to our results. We conclude with a summary or our contribution and a presentation of our perspectives.

2   *MAS*-Based Road Traffic Simulation Approaches

Multi-Agent Systems (*MAS*) allow the simulation of complex phenomena that cannot easily be described analytically. They are often based on the coordination and interaction of agents that lead to the emergence of the simulated phenomena [13–15]. The multi-agent approaches are well-suited to the applications such as road traffic simulation.

The main advantages of multi-agent models rely on the environment's dynamic modifications with a response time that is close to real time: preferences and characteristics of autonomous vehicles, appearance of vehicles (*e.g.*, buses, motorbikes, cars), pedestrians and the road signs (*e.g.*, stop signs, speed-limit signs). Moreover, the agents perceive local information (*i.e.*, geographically limited and may be incomplete). Finally, the traffic situation is, by nature, an open system (*i.e.*, the number of autonomous agents can vary during the simulation) in which the

---

[1] The French National Institute for Transport and Safety Research (ex-INRETS).

various entities (with different objectives) interact with each other. The situation is defined by multiple interactions between entities in their environment, which makes it possible to reproduce more realistic behaviors of human drivers. In fact, simulation conditions can be dynamically modified: the weather, the driving preferences of the human drivers, the characteristics of the autonomous agents (*e.g.*, cars, lorries, buses, pedestrians) and road equipment (*e.g.*, traffic signals, traffic signs).

Two research categories have emerged in terms of road traffic simulations using *MAS* [16,17]:

- the first approach defines organizational models to improve global problems, such as logistics and/or services [18,19],
- the second approach offers solutions for "local" traffic congestion problems. Congestion is a deteriorated state because all agents make "optimum" local decisions *a priori*, far from the global optimum. Several studies deal with these congestion problems:
  - A first category for this approach, is based on the optimization of global traffic [16,20–23].
  - A second category describes "profiles" for the different agents (for understanding and reproducing the human driving), analyses the impact of these profiles on the global traffic and extracts global information on the simulation (*e.g.*, statistics data concerning the average speed of vehicles, the number of accidents) to compare it to real observed data: the agents have specific profiles such as prudent or aggressive behaviors [24], predefined behaviors (normal, prudent or aggressive) with parameters (*e.g.*, inter-vehicular distance or acceleration-breaking characteristics) [25], or the agents may also have non normative behaviors (*i.e.*, not respecting the highway code, not breaking at a stop sign) [26].

Our approach is rather the reproducing of human driving than the optimization of global traffic. The paper aims to simulate the behavior related to the road space occupation, particularly in context of high traffic density in urban areas. The presence of road markings does not always prevent drivers to readapt the road space according to their goals and context. We can consider that each driver overloads the road structure defined by road marking by constructing his/her own ego-centered representation which meets his/her goals. The fact that users can define different ego-centered representations for the same "physical" configuration can be a source of conflict. To improve heterogeneous traffic simulation, we need to understand the behavior of the different types of drivers. There has been some empirical studies [27] which aim to understand the motorcycle behavior and the properties of mixed flow.

For a more realistic simulation, we need changing-lanes models taking into account this kind of phenomena. Several multi-agent traffic simulation approaches model changing-lanes mechanisms [3–5]. Hidas [3] introduces behavior heterogeneity through two kinds of behavior: aggressivity and courtesy. Dai and Li [5] consider not only the leader vehicle, but also the information from vehicles farther away in the lane changing process. However, in these work, the lanes used by drivers correspond to the physical lanes defined by the markings. These models do not consider the observed phenomena of road space occupation. The resulting simulations do not reproduce real situations. Furthermore, Fellendorf et al. [28] use VISSIM, which is a commercial simulation tool based on mathematical models, to describe the continuous lateral movement for the case of heterogeneous traffic situations. The driver chooses the lateral position where he has the maximum longitudinal time-to-collision. In our opinion, those parameters are not sufficient because the choice of the target lateral position is only based on an instantaneous evaluation.

Other work have proposed solutions for the two-wheels [6,7]. Lee et al. [7] rely on mathematical modeling; Bonte et al. [6] describe a *MAS* modeling. Bonte et al. [6] introduce therefore the concept of virtual lanes which are defined by measuring the free spaces on the road according to the position and width of vehicles. However, these models consider a systematic and geometric decomposition in virtual lanes of the space, this leads to a dynamic number of virtual lanes (it can be very high).

The next section presents the differences between two types of representation, *i.e.*, "ego-centered" or "allo-centered" models. Moreover, it describes the driver practices and shows that the driver practices may lead to real critical situations, difficult to design in simulations.

3   Driver Behavior and Environment Representation

3.1   Driving Psychology Studies and Driver Behavior

Driving a vehicle consists in carrying out a displacement in a constantly changing environment. To move, the drivers sustain a set of interactions described by the constraints of the other drivers' behavior, road infrastructure and regulation. A driver aims often at minimizing his/her travel time. So he/she tries to reach his psychological maximum speed, also called the desired speed. He/she thus considers his/her current state (*e.g.*, speed, position) and the various constraints imposed by his/her environment (*e.g.*, other vehicles, infrastructure).

The driver needs to have a representation of the environment around him/her to make his/her decision. Two theories were proposed to deal with the environment representation: allo-centered and ego-centered representations [29,30]. In ego-centered representations, spatial relations are generally directly related to the agent that builds a representation using a reference system with terms such as, for example, left, right, front, or back. When the context changes, all the spatial relations should be updated. Whereas, an allo-centered representation locates points within a framework external to the holder of the representation and independent of his/her position. Allo-centered representations are more stable but are more difficult to acquire. In addition, the number of the spatial relations is much higher since all the relations among different objects in the environment are considered.

The human driver "discovers" the situations as he/she moves. He/She needs to know what happens around him/her to make decisions (go straight, changing

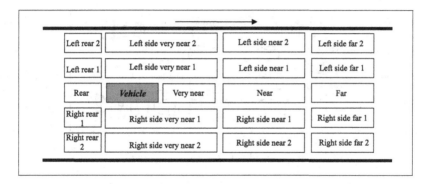

Left rear 2	Left side very near 2	Left side near 2	Left side far 2
Left rear 1	Left side very near 1	Left side near 1	Left side far 1
Rear	*Vehicle*	Very near	Near	Far
Right rear 1	Right side very near 1	Right side near 1	Right side far 1
Right rear 2	Right side very near 2	Right side near 2	Right side far 2

Fig. 1. The areas of perception of a simulated driver [4]

lane to the left or to the right). From this point of view, the ego-centered representations are more intuitive in the context of traffic simulation, for which we need to have a contextual and dynamic representation of what is happening around the agent. Furthermore, the ego-centered representation is suitable to dynamic contexts because the number of relations to update is lower than the number in the case of the allo-centered representation.

Saad [8] describes an ego-centered environment representation of the driver and considers that the space around the vehicle is the control field of the driver and can be divided into several sectors according to their location (front, back, left side, right side) and their proximity (very close, near, far, far away). This work focuses on the current lane of the vehicle, those on the right and left immediately adjacent lanes. For implementation reasons, and to deal with more road situations, El Hadouaj [4] extrapolates this same reasoning and adds non immediately adjacent lanes (right and left) to take into account the traffic on these lanes and allow the remove of blocking situations for highways over three lanes (Fig. 1). This solution described by El Hadouaj [4] is not generic because it does not address a number of situations where the favorable option requires more than two changing lanes such as highways with many lanes, a toll gate, etc. Furthermore, this representation is composed of physical lanes given by the road marking. Thus, it does not permit to identify free space on the road and cannot reproduce the observed practices of road space occupation in the simulation like the behavior of two-wheels vehicles.

El Hadouaj [4] chooses a pre-selection of all existing lanes to keep the most relevant one. But the choice of the retained lanes cannot allow to unblock all the road situations. Figure 2 illustrates a road with 4 lanes. For motorcycle a_1, it is blocked on its lane and the traffic situation is deadlocked on the two adjacent lanes to the left due to an accident. The last lane to the left has good features. In El Hadouaj's representation [4], the agent gets stuck on its lane when it could identify earlier that the traffic situation is better on the leftmost lane and anticipate the deadlock. It would be more interesting to explore the environment that is not directly adjacent and to identify areas that may be more advantageous, even if they are not directly reachable.

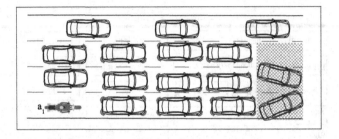

Fig. 2. Deadlock due to an accident

Moreover, Saad [8] describes some factors which are used by the driver in his/her decisions. The decisions of the driver rely on the properties of each zone, namely, their type in terms of infrastructure, the regulations governing them and the users' behavior. Further, psychological studies mention the concept of wall effect and highlight its impact on the driving speed as well as the driver lateral position on the lane. The wall effect may be related to infrastructure characteristics (*e.g.*, lane width, tunnel walls) or on the road context (*e.g.*, the effect of the presence of trucks on adjacent lanes, the speed variability of adjacent lanes) [9–12]. To summarize the results of these work, the driver speed is lower on the lane close to the tunnel wall than the other lanes, the narrower lanes generate lower speeds and vice versa and a wall effect may occur depending on road context (types of vehicles such as trucks, cars, buses).

3.2   Actual Practices of Drivers

The observation of actual practices shows that the drivers do not often respect the regulation in order to be more efficient, for example in terms of travel time (sometimes at a collective level but more often for personal gain). Drivers sometimes tend to readapt the road space by building their own representation of the environment which may not comply with the norms.

Figure 3 illustrates a traffic situation. Vehicle x is constrained by Vehicle y which is improperly parked. Two cases may be distinguished. In the first case, the driver has a "normative behavior" and chooses to make a changing lane (by using the physical lane) if it is possible. He/She will be constrained by Vehicle z which has a lower speed. In the second case, the driver observes the situation and finds that there is a free space on the road between Vehicles y and z. He/She chooses this emergent space. If Vehicle z is cooperative, he/she shifts to the left. We can consider that this kind of behavior is non normative.

We also observe the same behavior at intersections (Fig. 4). Vehicle z going straight, is constrained by two slower Vehicles x (turning to the right) and y (turning to the left). In this situation, the driver has not always a normative behavior. He/She tries to unblock his/her situation through a "virtual" lane formed by the space between x and y, especially as Vehicle y will generally tend to tighten to the left and Vehicle x to the right. Other examples of these

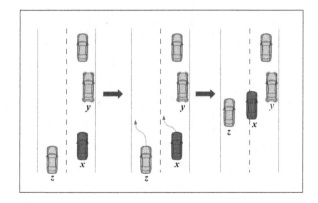

Fig. 3. Case of badly parked vehicle.

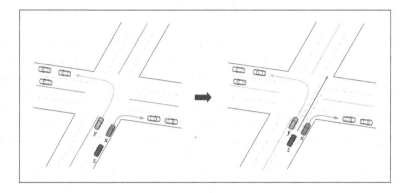

Fig. 4. Case of crossroad

practices concerning intersections and roundabouts have been already described in previous studies [31, 32].

Figure 5 illustrates a last example of observed practices in the reality: the motorcycles' moves are different from the other road users [33]. Due to their size, motorcycles occupy road space differently. The motorcycle drivers have non normative behaviors because they do not strictly follow the Highway Code. These behaviors (filtering maneuvers between vehicles) are forbidden by the Highway Code. Vehicles must drive on the physical lanes bounded by road marking or on lanes specifically dedicated to this type of vehicle. However these practices are tolerated in some countries and even are allowed in others.

The above examples show the fact that, in some situations, the driver chooses practices that are not necessarily conform to the regulations (a non normative behavior). These practices are related to a temporary re-adaptation of road space. Sometimes it is also the result of cooperation between individuals (case of emergency vehicles or motorcycles filtering along a car queue).

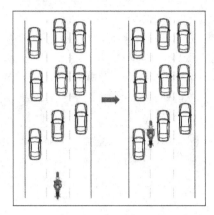

Fig. 5. Case of motorcycles

To summarize, our approach relies on the results of psychological studies. Moreover, we use the concept of virtual lanes introduced by Bonte et al. [6]. However, these authors consider a systematic and geometric decomposition in virtual lanes of the space; this can lead to a high number of virtual lanes. So, we introduce an ego-centered representation of the environment around the agent by selecting the lanes which represent the best alternatives (to the left and the right). Thus, our approach does not use a systematic and geometric breakdown of the space.

The following section describes the *ArchiSim* architecture which is based on a *MAS* approach for road traffic simulation.

4   Simulation Tool: *ArchiSim*

ArchiSim is a behavioral traffic simulation tool [1]. The latter uses a neat simulation of road traffic based on psychological researches on the driver behavior [8]. The traffic is considered as an emergent phenomenon resulting from the actions and interactions of the various road actors (*e.g.*, car drivers, pedestrians, road operators).

The core of the *ArchiSim* architecture (see Fig. 6) is a process capable of building, upon request, a symbolic description of the context of each agent. This "view server" contains all data related to the simulated environment as a description of the network, road equipment and the users evolving there. *ArchiSim* is a synchronous simulator, a simulation is divided into time steps, at each time step each agent indicates to the "view server" its new state "visible" by the others (position, speed, indicators status, etc.) and requests from the server the elements present in its perception field (adjustable in distance).

This process does not interfere in the decision-making of each agent: it is only responsible for delivering information. Each agent is autonomous: it has its own knowledge, goal and strategy to carry out its tasks and resolve any

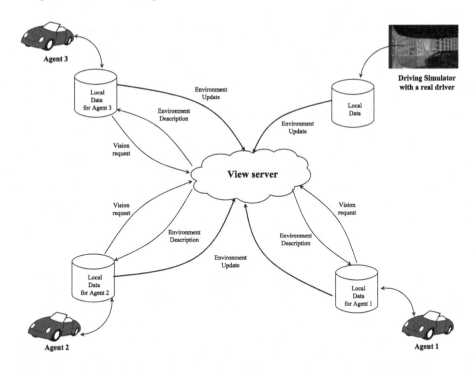

Fig. 6. *ArchiSim*architecture

conflict. It operates according to the scheme: perception, decision and action. In fact, the decision is based on the agent's knowledge of the context in which it operates. To move on the road and to adapt to the evolution of the context, each agent needs to perceive the different elements of its environment, elements which are provided by the "view server". It evaluates the parameters related to the context from its current situation and considering the probable evolution of the context. It therefore builds an ego-centered vision of its environment, the perceived elements being located regarding to itself (same road, same lane, forward, backward, relative distance). An ego-centered representation is a vision that considers the short/medium term goals of the driver. For example, the driver focuses differently on the various branches of an intersection if she/he plans to go straight, left or right.

*ArchiSim*provides a set of tools (Fig. 7) to facilitate the experiments requiring the traffic production [34]:

– Network setting up (*Wr2*): *Wr2* allows to generate automatically different files describing the roads and intersections of the network: axial kilometer points, the network structure (graph), building of roads (boundaries and side-walks), the markings of the various sections, the allocation of lanes, the signals (static and dynamic panels, light controllers).

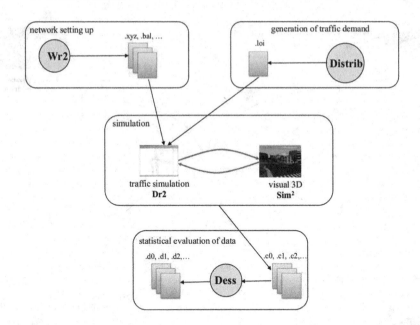

Fig. 7. *ArchiSim*tools

- Generation of traffic demand (*Distrib*): *Distrib* produces a traffic with different vehicles. Each vehicle is described by the time step when it must be created, its location on the network at its inception, its acceleration and initial speed, its itinerary, the various behavioral parameters (*e.g.*, experience, distance to the regulation).
- The core of the simulation and traffic model (*Dr2*): *Dr2* takes as input the network description produced by *Wr2* and the traffic demand generated by *distrib*. It provides a 2D top view of the simulated network. Note that a 3D visual also exists (*sim2*) which allows human drivers, using driving simulators, to participate to the simulation. Driving simulation *sim2* can be executed in parallel with *Dr2* and communications between the two software enable a 3D visualization of the traffic scene.
- Data processing (*Dess*): The user can put a set of virtual sensors on the network. *Dess* permits to process the files created by the virtual sensors in order to aggregate data.

Note that our model deals with the impact of the behavior heterogeneity (individual behavior, vehicle type, etc.) in the decision making for the different agents. This heterogeneity leads to different representations of the same environment. The individual characteristics of drivers are specified with the *distrib*.

We present in the next section, our approach, called "Ego-centered Environment Representation Model" (ECERM), which has been implemented in the *ArchiSim*tool. Our representation of the environment changes from the physical road based on different lanes to virtual lanes. We give the different steps for building the different agents' views.

5   The Ego-Centered Environment Representation (ECER) Model

5.1   Definition of the Ego-Centered Environment Representation

In a given traffic situation, the driver has the choice between staying on his/her lane and adapting to the constraints or changing lane.

The agent needs to build an ego-centered representation of the world around it. We made the assumption that the world in which the agent evolves is not defined only by the physical lanes but it can also be built by overloading the existing structure. Therefore, we propose to define driver agent field of control through the concept of virtual lanes using only five virtual lanes (and not a systematic and geometric breakdown of all road spaces):

– the current lane of the agent,
– two adjacent lanes (right and left) for which a geometry is defined,
– two lanes which represent the existence of a lane "reachable" to the right or left, beyond the adjacent lanes. These lanes are not necessarily doubly adjacent (adjacent to adjacent lanes). They indicate lanes that are reachable by a series of changing lane maneuvers, such as a favorable option reachable at the cost of changing lane sometimes unfavorable.

5.2   New Approach Based on an Ego-Centered Environment Representation Using Virtual Lanes

The virtual lanes are necessary for the design of our ego-centered environment representation. Figure 8 describes the four steps useful for our approach:

1. The agent breaks down the roadway on occupied and empty lanes depending on the width and position of the perceived vehicles (for a given distance of vision).

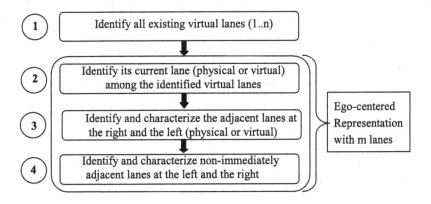

Fig. 8. Construction steps of the ego-centered environment representation

2. The agent determines its own lane which may correspond to a virtual lane (if it presents good properties, see the following section) or a physical lane.
3. The choice between virtual and physical lanes is made by evaluating the virtual lane according to the lane properties and the individual characteristics of the agent (to evaluate the virtual lane attractiveness in terms of gain).
4. The agent chooses the first lane whose characteristics are better than those of the current lane.

5.3 Lane Characteristics

We assume that the estimation of expected gain depends on the flow characteristics, the walls effect of the target lane and the agents' individual characteristics.

The flow characteristics of each lane can be translated in terms of: the lane length (*i.e.*, the depth of the lane), the lane density, the average speed for vehicles and the speed standard deviation which measures the speed distribution on the lane and therefore the stability of traffic in terms of speed (a high standard deviation would mean that traffic is not stable and therefore not predictable). Regarding to wall effect, we keep these properties for our model:

- Speed: we consider an average speed of vehicles that are on the adjacent lanes defining the wall, or null if it is a roadside.
- Stability: the stability of the walls is given by the difference between the average speeds of each wall. We postulate that the more the speeds of right and left walls tend to be identical the more the wall effect is considered as stable.
- Proximity: the proximity reflects the available space between the vehicle and the edges. This area affects the driver speed on the lane (more the space is reduced, more the vehicle speed decreases).

We note that the impact of the lanes' characteristics varies according to the criteria within and between individuals. Characteristics are related to:

- The distance that the driver accepts in relation to regulation, more this distance is smaller, more the choice of using a prohibited lane has a significant cost. This distance also varies depending on the types of users (traffic interqueues is often tolerated for motorcycles and it is also a practice for emergency vehicles).
- The social acceptance of the virtual lane use (concept of tolerance for the others).

5.4 Lane Evaluation

To build the ego-centered environment representation based on the virtual lanes concept, the agent evaluates the identified virtual lanes in order to choose those which represent the best alternatives to the left and right. The evaluation mechanism is based on the lanes' properties identified above. Each agent has a choice between staying on its own lane or switch to another one. This evaluation is done through a gain function that compares the agent current velocity and the target lane expected velocity. The gain function is given by the difference between the two velocities[2]:

$$G\left(vc_{a_i}, v_{a_i}\left(l_j\right)\right) = t_{a_i} \times v_{a_i}\left(l_j\right) - vc_{a_i}$$

where a_i is Agent i, l_j is lane j, $v_{a_i}\left(l_j\right)$ is the expected agent velocity on lane l_j, vc_{a_i} is the agent current velocity and $t_{a_i} \in [0,1]$ reflects the social acceptance of the filtering practice that differs according to the vehicle type (*e.g.*, motorcycle, car, bus). $v_{a_i}\left(l_j\right)$[3] depends on the following parameters:

- $f_{a_i}\left(l_j\right)$: reflects the flow characteristics of lane l_j and depends on the lane density and on the lane average speed,
- $g_{a_i}\left(l_j\right)$: reflects the wall effect of lane l_j and depends on the closeness of the walls and their stability in terms of speed,
- $h_{a_i}\left(l_j\right)$: is related to the individual characteristics of each agent and translates its distance to the regulation (normative and non normative behavior).

The evaluation function is positive when the target lane is relevant for the agent (in terms of speed). This estimation takes into account the lane characteristics (*e.g.*, width, depth), the wall characteristics (stability, proximity) and the individual agent characteristics, especially the distance to the norm.

With the generalization of the virtual lanes use as well as the enrichment of lanes' properties, we expect that the alternative to choose a virtual lane (inter-queues) will not be systematic and particularly for cars, trucks and buses, where "tolerance" associated with the use of such lanes is low as well as the gain in terms of travel time. For these users, it will be more favoured in case of specific events (presence of a vehicle badly parked, emergency vehicle). The filtering maneuver will be more reserved to two-wheels because of greater tolerance and a significant gain in terms of travel time. The proposed solution is also expected to improve the validity of the model for situations with an important number of lanes, in particular the consideration of "complex" tolls.

The next section describes two scenarios illustrating the validity of our approach and describes the realized experiments.

6   Experiments and Results

Two scenarios where the agents are in situations of traffic suitable for observing the desired behavior are described to evaluate the individual behaviors of the agents in terms of space occupation. For these scenarios, we compare the agent behavior in the benchmark case (without virtual lanes) and in our model. We also consider the impact of the behavior in terms of heterogeneity (individual behaviors, vehicle types, etc.).

---

[2] The choice of this function is completely empirical; we have chosen the parameters that affect the behavior of the agent based on psychological studies.

[3] The expected agent velocity on lane l_j is given by the weighted sum of the parameters mentioned below.

6.1   Scenario 1: Behavior of Motorcycle Overtaking Vehicles

This first scenario aims at verifying the filtering behavior of motorcycles (Fig. 9). The situation describes a road with 2 physical lanes with Motorcycle 1 and two slower vehicles (2 and 3). We focus on the behavior of Vehicle 1 (a motorcycle) for the two approaches, *i.e.*, the benchmark model and in ECER model (our model).

We can observe that, in the case of the benchmark model, the motorcycle still behind Vehicle 3 although the latter has a slower speed. In this case, the motorcycle uses the physical lanes and even if the driver makes a lane change to Lane 1, it still constrained by Vehicle 2. Figure 9 shows also that in the case of our model (ECER model) the agent driving two-wheels chooses at the time step 11 to fit and filter between the two vehicles.

The behavior of motorcycle results from the ego-centered environment representation based on virtual lanes. The motorcycle agent detects the possibility of the emergent virtual lane (between the two vehicles). This kind of behavior does not appear in the reference model (which has also been implemented in *ArchiSim*, without our model), as the lanes correspond to the physical ones.

For the two cases, we compare the velocity and the lateral position of Vehicle 1. In the benchmark model, the motorcycle does not filter between cars. In our model, the motorcycle driver evaluates the virtual lane (between the two cars) and chooses it. This behavior allows it to reach its desired velocity (Fig. 10).

We also compare the car and the motorcycle behaviors to the ECER model. We can observe that in the same configuration, the motorcycle driver chooses the

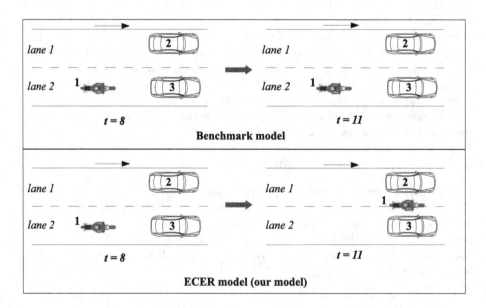

Fig. 9. Comparing the positioning behavior in benchmark and ECER models

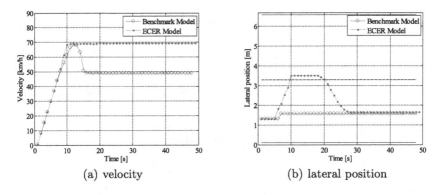

Fig. 10. Results for Vehicle 1

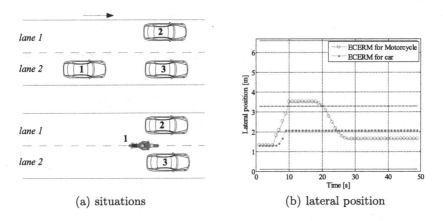

Fig. 11. Comparing the lateral position in ECER model for car and motorcycle

virtual lane whereas the car driver stays behind Vehicle 3 (Fig. 11). The lateral positions shown on the right side of the figure illustrate these behaviors.

6.2   Scenario 2: Influence of Normative vs Non Normative Behaviors

The second scenario study the impact of individual characteristics such as the distance to the regulation. We change the individual characteristics of the agent in order to have two different behaviors: normative and non normative. In this case, Agents 1 and 3 are cars and Agent 2 is a motorcycle (Fig. 12).

In the first case, the agent has a normative behavior. It acts as in the benchmark model and stays behind Vehicle 3 even if it has a lower speed because the alternative of virtual lane is costly. In the second case, the agent has a non normative behavior, it chooses the virtual lane because it may enable it to have a higher speed. With these two cases, we consider two extreme classes of behaviors. In our model, the variety of behaviors between those two extremes may be produced.

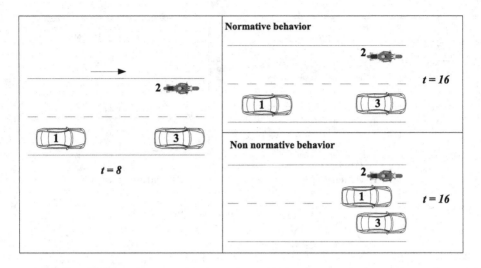

Fig. 12. The car behavior (normative and non normative) with ECER model

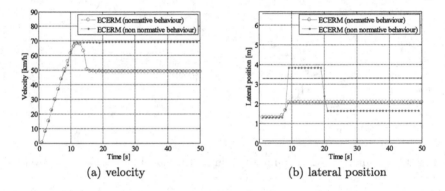

(a) velocity   (b) lateral position

Fig. 13. Results for Vehicle 1

The velocity curves (Fig. 13) show that with normative behavior, the agent accelerates to reach its desired speed but slows quickly to adapt to the vehicle in front which is slower. It is not the case with the non normative behavior where the agent accelerates until its desired speed and can keep this speed.

To summarize, our model takes into account the fact that the filtering practice is more tolerated for the motorcycle than for the car. We can observe that the car drivers choose the alternative of a virtual lane only if the agent has a non normative behavior (extreme case). The choice of virtual lanes is not systematic; it depends on the lanes' characteristics as well as the vehicle characteristics and the agent individual characteristics.

Simulations in a road situation with real heavy traffic have been studied. We compare the results of our model with real ones. The details of these experiments can be found in [35]. Those results show that the two-wheeled vehicle took twice less time than cars to make the same journey. Due to their size, two-wheeled vehicles use virtual lanes more frequently than other vehicles types. Their travel time is shorter than for the travel time of cars. We can thus conclude that our model considers that filtering is better tolerated in two-wheeled drivers than in car drivers.

7 Conclusion

Our work intends to extend the validity of traffic simulation in urban and sub-urban areas, with a better consideration of the heterogeneity of the vehicles and driver behaviors in terms of anticipation positioning on the lane and space occupation.

To reach this objective, we considered that the *MAS* approaches may allow us to produce more realistic behaviors at individual and collective levels. In this case, the existing studies applied to road traffic have shown that one of the main difficulties relies on the re-organization of road space occupation. Each driver builds his/her own representation for the same road configuration. In existing driving psychology studies, two theories were proposed to deal with the environment representation: allo-centered and ego-centered representations. The ego-centered representation is closer to *MAS* principles. Moreover, three examples (*i.e.*, vehicle improperly parked, situation related to a typical intersection and situation dependent to motorcycles' drivers), which have been described in this paper, outlined that the real practices of drivers do not always respect the normative behavior (*i.e.*, the "normal" driving): these practices define a temporary re-adaptation of road space.

To support our analysis, *ArchiSim*architecture has been described. *ArchiSim*is a road traffic tool based on psychological researches on the driver behavior. This tool allows to produce the actions and interactions for various agents (*e.g.*, vehicles, motorcycles), each agent having its own characteristics and goals.

We proposed to use for each agent, the concept of virtual lanes (*i.e.*, a solution to re-define the road occupation space) coupled with an ego-centered representation of the traffic situation. Our solution is based on four steps: (1) identify all the virtual lanes, (2) identify the current lane among these different virtual lanes, (3) identify and characterize the adjacent lanes, (4) identify and characterize not immediately adjacent lanes at the left and the right. To characterize the choice of a lane, the agents evaluate the different interesting lanes, w.r.t. speeds of vehicle, width and depth of lanes, conformity to norm regulation, etc.

Finally, we validate some individual behaviors for specific situations. Two experiments were thus explained. The first experiment deals with the behavior of motorcycle, which overtakes vehicles. The reference model leads the motorcycle to stay behind the vehicles. However, the ego-centered representation model allows the motorcycle to overtake the vehicles and to reach its desired speed (the car allows to stay behind the other vehicles, because the virtual lane is not sufficient to overtake). These results are similar to real practices. The second experiment aims at improving the existing traffic models by the simulation of normative and non normative behaviors. The agent may re-define the road space occupation and its representation may be also modified according to the type of behavior. The vehicle may move between a motorcycle and a vehicle in a situation where the vehicle has a non normative behavior (the width of virtual lane is sufficient to overtake them).

We investigated that the function determining the gain for virtual lane may be evaluated more precisely. A study concerning its impacts seems necessary. Furthermore, it is interesting to compare our model to other general approaches for such heterogeneous contexts. However, to compare different models, we need to implement them in the same tool and to define experiment in the same context. It is an interesting issue but it is time consuming.

Acknowledgements. This research was partially funded by the French Ministry of Education, Research and Technology, the Nord/Pas-de-Calais Region, the CNRS, the International Campus on Safety and Intermodality in Transportation (CISIT). We would like also to thank the anonymous reviewers for their comments.

References

1. Espié, S.: Archisim, multi-actor parallel architecture for traffic simulation. In: Proceedings of the Second World Congress on Intelligent Transport Systems, Yokohama, vol. IV (1995)
2. Ksontini, F., Espié, S., Guessoum, Z., Mandiau, R.: Traffic behavioral simulation in urban and suburban - representation of the drivers' environment. In: Demazeau, Y., Müller, J.P., Rodríguez, J.M.C., Pérez, J.B. (eds.) Advances on PAAMS. AISC, vol. 155, pp. 115–125. Springer, Heidelberg (2012)
3. Hidas, P.: Modelling lane changing and merging in microscopic traffic simulation. Transp. Res. Part C: Emerg. Technol. **5–6**, 351–371 (2002)
4. El Hadouaj, S.: Conception de comportements de résolution de conflits et de coordination: application à une simulation multi-agent du trafic routier (2004)
5. Dai, J., Li, X.: Multi-agent systems for simulating traffic behaviors. Chin. Sci. Bull. **55**, 293–300 (2010)
6. Bonte, L., Espié, S., Mathieu, P.: Modélisation et simulation des usagers deux-roues motorisés dans archisim. In: Actes des14e Journées Francophones sur les Systèmes Multi-Agents (JFSMA'06), pp. 31–44 (2006)
7. Lee, T., Polak, J., Bell, M.: New approach to modeling mixed traffic containing motorcycles in urban areas. Transp. Res. Rec. **2140**, 195–205 (2009)
8. Saad, F.: In-depth analysis of interactions between drivers and the road environment: contribution of on-board observations and subsequent verbal report. In: Proceedings of the 4th Workshop of ICTCT, University of Lund (1992)
9. Tornros, J.: Driving behavior in a real and a simulated tunnel - a validation study. Accid. Anal. Prev. **30**, 497–503 (1998)
10. Fitzpatrick, K., Carlson, P., Brewer, M., Wooldridge, M.: Design factors that affect driver speed on suburban streets. Transp. Res. Rec. **1751**, 18–25 (2001)
11. Lewis-Evans, B., Charlton, S.G.: Explicit and implicit processes in behavioural adaptation to road width. Accid. Anal. Prev. **38**(3), 610–617 (2006)
12. Schramm, A., Rakotonirainy, A.: The effect of traffic lane width on the safety of cyclists in urban areas. J. Australisian Coll. Road Saf. **21**(2), 43–49 (2010)

13. Ferber, J.: Multi-Agent Systems: An Introduction to Distributed Artificial Intelligence. Addison Wesley, Reading (1999)
14. Guessoum, Z., Mandiau, R.: Modèles multi-agents pour des environnements complexes. In: Numéro spécial de la Revue Française d'Intelligence Artificielle (RIA), vol. 21. Hermes (2008)
15. Weiss, G.: Multiagent Systems: A Modern Approach to Distributed Artificial Intelligence. MIT Press, Cambridge (2000)
16. Meir, R., Rosenschein, J.: A game theoretic approach to leasing agreements can reduce congestion. In: 6th Workshop on Agents in Traffic and Transportation, Co-Located with the 8th International Joint Conference on Autonomous Agents and Multi-Agent Systems (ATT@AAMAS 2010), Toronto, Canada, pp. 21–27 (2010)
17. Kubicki, S., Lebrun, Y., Lepreux, S., Adam, E., Kolski, C., Mandiau, R.: Simulation in contexts involving an interactive table and tangible objects. Simul. Model. Pract. Theory 31, 116–131 (2013)
18. Davidsson, P., Henesey, H., Ramstedt, L., Tornquist, J., Wernstedt, F.: Agent-based approaches to transport logistics. In: Application of Agent Technology Traffic and Transportation (ATT 2005), pp. 1–16 (2005)
19. Zeddini, B., Zargayouna, M., Yassine, A.: Space and space-time organization model for the dynamic vrptw. In: 6th Workshop on Agents in Traffic and Transportation, Co-Located with the 8th International Joint Conference on Autonomous Agents and Multi-Agent Systems (AAMAS 2010), Toronto, Canada, pp. 21–27 (2010)
20. Dressner, K., Stone, P.: A multi-agent approach to autonomous intersection management. J. Artif. Intell. Res. 31, 591–656 (2008)
21. Vasirani, M., Ossowski, S.: A computational market for distributed control of urban road traffic systems. IEEE Trans. Intell. Transp. Syst. 12(2), 313–321 (2011)
22. Bazzan, A.L.C., Wahle, J., Klügl, F.: Agents in traffic modelling - from reactive to social behaviour. In: Burgard, W., Christaller, T., Cremers, A.B. (eds.) KI 1999. LNCS (LNAI), vol. 1701, pp. 303–306. Springer, Heidelberg (1999)
23. Bazzan, A.: A distributed approach for coordination of traffic signal agents. Auton. Agent. Multi-Agent Syst. 10(1), 131–164 (2005)
24. Ehlert, P.A., Rothkrantz, L.J.: Microscopic traffic simulation with reactive driving agents. In: IEEE Intelligent Transportation Systems Conference Proceedings, pp. 861–866 (2001)
25. Paruchuri, P., Pullalarevu, A.R., Karlapalem, K.: Multi agent simulation of unorganized traffic. In: Proceedings of the First International Joint Conference on Autonomous Agents and Multiagent Systems: Part 1, AAMAS '02, pp. 176–183. ACM, New York (2002)
26. Doniec, A., Mandiau, R., Piechowiak, S., Espié, S.: Controlling non-normative behaviors by anticipation for autonomous agents. Web Intell. Agent Syst. 6(1), 29–42 (2008)
27. Minh, C.C., Sano, K., Matsumoto, S.: The speed, flow and headway analyses of motorcycle traffic. J. Eastern Asia Soc. Transp. Stud. 6, 1496–1508 (2005)
28. Fellendorf, M., Vortisch, P.: Microscopic traffic flow simulator vissim. Fundam. Traffic Simul. Int. Ser. Oper. Res. Manage. Sci. 145, 63–93 (2010)
29. Cohn, A., Renz, J.: Qualitative spatial representation and reasoning. In: van Harmelen, F., Porter, B. (eds.) Handbook of Knowledge Representation, vol. 3, pp. 551–596. Elsevier, Amsterdam (2008)
30. Wang, H., Kearney, J. Cremer, J., Willemsen, P.: Steering behaviors for autonomous vehicles in virtual environments. In: Proceedings of the IEEE Virtual Reality Conference, Bonn, Germany, pp. 155–162 (2005)

31. Doniec, A., Espié, S., Mandiau, R., Piechowiak, S.: Non-normative behaviour in multi-agent system: some experiments in traffic simulation. In: Proceedings of the 2006 IEEE/WIC/ACM International Conference on Intelligent Agent Technology (IAT), Hong Kong, China, 18–22 December 2006. IEEE Computer Society, pp. 30–36 (2006)

32. Doniec, A., Mandiau, R., Piechowiak, S., Espié, S.: Anticipation based on constraint processing in a multi-agent context. J. Auton. Agent. Multi-Agent Syst. (JAAMAS) **17**(2), 339–361 (2008)

33. Aupetit, S., Espié, S.: Analyse ergonomique de l'activité de conduite moto lors de la pratique de l'inter-files en région parisienne. Activités **9**, 48–70 (2012)

34. Ksontini, F., Espié, S., Guessoum, Z., Mandiau, R.: A driver ego-centered environment representation in traffic behavioral simulation. In: Demazeau, Y., Müller, J.P., Rodríguez, J.M.C., Pérez, J.B. (eds.) Advances on PAAMS. AISC, vol. 155, pp. 249–254. Springer, Heidelberg (2012)

35. Ksontini, F., Guessoum, Z., Mandiau, R., Espié, S.: Using ego-centered affordances in multi-agent traffic simulation. In Gini, M.L., Shehory, O., Ito, T., Jonker, C.M., eds.: International Conference on Autonomous Agents and Multi-Agent Systems, AAMAS '13, Saint Paul, MN, USA, 6–10 May 2013, IFAAMAS, pp. 151–158 (2013)

Using LCS to Exploit Order Book Data
in Artificial Markets

Philippe Mathieu and Yann Secq[✉]

Laboratoire d'Informatique Fondamentale de Lille (UMR CNRS 8022),
Université Lille 1, Villeneuve-d'Ascq, France
{philippe.mathieu,yann.secq}@univ-lille1.fr
http://www.lifl.fr/SMAC

Abstract. In the study of financial phenomena, multi-agent market order-driven simulators are tools that can effectively test different economic assumptions. Many studies have focused on the analysis of adaptive learning agents carrying on prices. But the prices are a consequence of the matching orders. Reasoning about orders should help to anticipate future prices.

While it is easy to populate these virtual worlds with agents analyzing "simple" prices shapes (rising or falling, moving averages, ...), it is nevertheless necessary to study the phenomena of rationality and influence between agents, which requires the use of adaptive agents that can learn from their environment. Several authors have obviously already used adaptive techniques but mainly by taking into account prices historical. But prices are only consequences of orders, thus reasoning about orders should provide a step ahead in the deductive process.

In this article, we show how to leverage information from the order books such as the best limits, the bid-ask spread or waiting cash to adapt more effectively to market offerings. Like B. Arthur, we use learning classifier systems and show how to adapt them to a multi-agent system.

Keywords: Agent based computational economics · Artificial stock market · Market microstructure · Learning classifier systems · Multi-agent simulation

1 Introduction

In recent years, advances in computer research have provided powerful tools for studying complex economic systems. Individual-based approaches, for their benefits and the level of detail they provide, are becoming increasingly popular within industries and even for policy makers. It is now possible to simulate complex economic systems to study the effects of new regulations, or the influence of new policies at the individual and not only at the group level.

Among these economic systems, artificial financial markets now offer a credible alternative to mathematical finance and econometrical finance. Thanks to multi-agent systems, decision-making as the actions taken can be individualized, macroscopic phenomena becomes consequences of microscopic interactions.

© Springer-Verlag Berlin Heidelberg 2014
R. Kowalczyk et al. (Eds.): TCCI XV, LNCS 8670, pp. 69–88, 2014.
DOI: 10.1007/978-3-662-44750-5_4

1.1   An Artificiel Agent-Based Stock Market

One can found many artificial markets in the literature, but they are mostly built at the macroscopic level, with prices defined through equations [1,15]. Agents reason on price alone and send simple signals to the market to buy or sell an asset. The market contains no order book and sets the price only on the differential between asks and bids signals. This approach ignores the complexity of real markets and is insufficient to test sophisticated behavioural assumptions like reasoning on types of orders, prices or quantities. Thus, to evaluate possible consequences of regulatory rules on individuals, or to study societies social welfares [7] or even speculators influences within traders population, a granularity at the individual is required.

The ATOM multi-agent platform [8] implements an order-driven market that reproduce the main features of the microstructure of marketplaces like *EuroNEXT-NYSE* , including its system of double book of orders for each asset. In ATOM, agents rely on their own strategies to send orders on different assets. ATOM is built on classic agent design-patterns [17] to allow the enforcement of equity between agents and to conduct experimentations at several scales: from an intra-day level (intraday) by reasoning on each fixed price to multi-day scale (extraday) by reasoning only on the closing prices. Although ATOM supports multi-assets negotiation, this article focuses on the reasoning on a single order book. Unlike macroscopic market models, agents do not emit a simple signal to buy or sell but can send real orders consistent with those allowed on *EuroNEXT*. However, in this paper, we limit ourselves to the two most common types: limit and market orders.

In this paper, we use only the best known and most widely used order, the `LimitOrder` which is defined by:

- the order issuer,
- the asset's name to be exchanged,
- the desired direction (bid or sell),
- the number of asset to be exchanged,
- the price limit (i.e. the maximum accepted price for a purchase order and the minimum accepted price for a sell order).

By its relevance to the real market mechanisms, ATOM provides access to numerous data. Among these, there is the price history (Table 2), but also and especially the orders history (Table 1). Double order books that reflect the state of the offer at time t is also accessible, with information on all orders its contains (Table 3).

These information are significant and allow a measure definition of how easily an agent will find a counterpart to his orders (so called *market liquidity*) or detect an imbalance between supply and demand that suggests a future price curve slope. Historically, economic theories claimed the importance to take advantage of these information [4,10,14], but so far nobody to our knowledge had highlighted this fact experimentally.

Table 1. Chronological history of sent orders.

Id.	Sender	Direction	Price	Quantity
o1	Agent2	sell	111.5	8
o2	Agent1	sell	111.1	10
o3	Agent1	bid	110.6	6
o4	Agent3	sell	111.0	10
o5	Agent1	sell	110.9	8
o6	Agent2	bid	110.9	8
o7	Agent2	bid	111.0	7
o8	Agent4	bid	110.9	2
o9	Agent3	sell	110.9	2
o10	Agent4	bid	110.8	7
o11	Agent2	sell	110.8	5

Table 2. Chronological prices history, along with quantities exchanged and agents involved in the transaction.

Price	Quantity	Bid order	Sell order
110.9	8	o6	o5
111.0	7	o7	o4
110.9	2	o8	o9
110.8	5	o10	o11

Table 3. Order book state after matching orders from Table 1.

Direction	Order	Sender	Quantity	Price
Vente	o1	Ag2	8	111.5
	o2	Ag1	10	111.1
	o4	Ag3	3	111.0
bid-ask spread $\updownarrow$				
Achat	o10	Ag4	2	110.8
	o3	Ag1	6	110.6

This paper main purpose is to demonstrate that it is possible to design trading behaviours that take advantage of all these information and offer a most efficient and effective trading behaviour.

1.2   Learning Trading Agents

As multi-agent systems, machine learning techniques have experienced a real boom in recent years. Main learning families (supervised, unsupervised, reinforcement) are based on different algorithms, the best known and most used are probably [12]:

- genetic algorithms,
- neural networks,
- Bayesian networks,
- support vector machines.

These techniques have been used in artificial markets with more or less success [2,11]. However, they all suffer the same explanatory default: once learning is achieved, it is difficult to understand why decisions are made, to highlight the cause of the outbreak of specific behaviors, and thus to avoid biases in learning contexts.

Different learning agent types have appeared within the literature in recent years. Even if these agents are designed to perform well on artificial markets, theirs learning is only focused on past prices to predict the possible future prices evolution.

As B. Arthur [1], we have chosen another learning technique that may be less common, but is much better adapted to the need of explanation: classifier systems or *Learning Classifier Systems* (LCS) [6]. These systems use a population of binary rules set by the designer that a reinforcement algorithm sort and that can possibly be modified by a genetic algorithm. Other techniques, such as those mentioned above would probably also work but our purpose here is not to make a comparison, but above all to show how to design an adaptive model and yet explanatory using the data set of one market to obtain varied and relevant behaviours.

One of the first artificial market, the SF-ASM (*Santa Fe Artificial Stock Market*, [1,15]), already used LCS for its reasoning agents, but this market is equational, and thus its agents could only take into account price history.

We show in this paper how to take advantage not only of past prices, but of all the available information in an order-driven multi-agent stock market simulator.

1.3 LCS Solely Based on Prices

Before discussing the overall complexity of a stock market, let's first describe the principle of a classical classifier system [1,9] by showing its usage in a macroscopic context where agents only study past prices to define their orders.

A LCS is initialized with a set of conditions on market state called indicators. These are the "sensors" used by the LCS to perceive market dynamic. Therewith, it is possible to derive a binary sequence whose length is equal the number of indicators used to characterize the current market state. Table 4 shows some indicators examples that can be used.

The first indicator Ind1 is satisfied if the current price is higher than the previous price. Ind2 is satisfied if the current price is higher than the average of the last 5 price, while ind3 is satisfied when the current price is less than 100.

These indicators being fixed, each LCS has a set of rules (or classifier) consisting of a triplet *(condition, score, action)*

Table 4. Technical indicators based on past prices

Market indicator name	Market indicator definition
Ind1	$p_t > p_{t-1}$
Ind2	$p_t > 1/5 \times \sum_{i=t-1}^{t-5} p_i$
Ind3	$p_t < 100$

Table 5. A LCS example using 5 rules

Rule	State S	Action A	Score F
R1	#10	bid	5
R2	1#0	sell	18
R3	00#	bid	12
R4	110	hold	4
R5	#11	sell	9

- the state S of a rule is a sequence of *trit* (***tr**inary dig**its***) that determines whether the rule can be activated given the current situation. These *trit* can be 0, 1 or #. In this sequence, each *trit* is an indicator. If a *trit* is 1 (resp. 0), the corresponding indicator must be *true* (resp. *false*) to enable the rule. The sign # means that the indicator is not be taken into account for rule activation.
- the ability to score F is the confidence we can have in this rule, based on its previous success prediction. The higher it is, the better is the rule.
- A action (bid or sell) to be performed if the rule is triggered. This choice is equivalent to a prediction on prices. Deciding to buy when we predict that prices will rise, and deciding to sell when we believe that prices will fall.

An LCS can have at most 3^n rules, where n is the number of indicators. Table 5 is an example of LCS with 5 rules using three indicators. For example, the first rule R1 can be activated if the current market situation is the state 010 or 110. This rule allows the LCS to select a purchase order, and his score is currently 5.

The LCS works as follows: every time he has to make a decision, it selects a rule among the activated rules with a probability proportional to the score of each rule. Then, the LCS performs the action associated with the selected rule (sending a signal to buy or sell). In the next activation of the LCS, the score of each activated rule will be corrected up or down depending on the accuracy of the prediction made.

In addition to the reinforcement system, it is common for a LCS to use also a genetic algorithm to renew its set of rules during simulation. It is applied regularly on the set of rules, using the score of each rule to produce natural selection within the classifier system. Rules whose score is below a threshold are eliminated, we cross and mute the best rules to regenerate the population.

The rules are not activated over time (due to conflicting indicators) are automatically eliminated. The combination of system update scores and genetic algorithm allows the LCS to achieve an effective learning.

There are other learning agents derived from LCS systems. For example, it is possible to make a social learning (information and learning are shared by multiple agents), or perform a hierarchical learning process multi-agents (HXCS [20]). However, LCS agents used here are agents of type XCS (*eXtended Classifier System*, [5,19]) realizing a simple reinforcement learning by updating the score of each rule.

2   An Order Based LCS

Adapting an LCS to an order-driven market poses several problems: temporal references within indicators may have different interpretations, in an order-driven market defining the direction (bid or sell) alone is not sufficient because it is necessary to also produce a proposal price and quantity, and finally, the agents must take advantage of information from pending orders contained in order books.

2.1   LCS and Temporal References

By adapting a classifier agent to a multi-agent system, the questions of the unity of time should be considered, because several indicators rely on this notion. Indeed, all agents can express themselves during a time step (in real time or turn to speak), but not always in the same order, and several prices may be fixed within one simulation step.

For example, if we want to know if the price has increased or decreased, the condition $p_t > p_{t-1}$ may not have the same meaning for each agent. In fact, agents have their own rhythms to place an order and a reference to past prices can be absolute or relative.

More precisely, to check the condition $p_t > p_{t-1}$, the agent can consider that t refers to either the known sequence of events by the market (the agent compares the current price at the last price fixed by the market), or the sequence of events known by the agent (the agent compares the current price at which prevailed the last time he made a decision, knowing that many prices have since been fixed).

We choose in this study to consider that the time is "the agent time", that is to say the sequence of events known by the agent, because it allows agents to reason about values spaced in time according to their need.

2.2   LCS and Order-Driven Stock Markets

When an agent send an order to the market, the direction of the order is determined by enabled rule, but it is also necessary set a price limit and a quantity. For a fair comparison between agents, we propose to use the same policy of pricing and quantity for all agents.

For pricing strategies, we have decided to rely on the best prices contained in the order book. For example, for the order book presented within Table 3, the values are:

$P_{BestBid} = 110.8$ and $P_{BestAsk} = 111.0$

Using these best prices, we propose two different strategies:

- fixing the price to place the order at the top of the order book:

  $P_{Bid} = P_{BestBid} + \varepsilon$

  $P_{Ask} = P_{BestAsk} - \varepsilon$

- launch an order that will be immediately matched (at least in part), with a price equal to the best opposite order:

  $P_{Bid} = P_{BestBid}$

  $P_{Ask} = P_{BestAsk}$

Pour fixer la quantit, nous proposons deux stratgies: soit une quantit constante $(Q = k_c)$, soit une quantit proportionnelle au score de la rgle active $(Q = k_p F)$.

By combining these two types of strategy, it is possible to design four different policies, we compare them in Sect. 4. Thus, the agent LCS allows him to determine the direction while policies set a price and a quantity. An agent with these two elements is able to send orders to the market.

2.3   Leveraging Order Books State Information

To show that the learning agents can be improved by taking into account market microstructure, we propose to give to previous agents the ability to access order books information. To this aim, we add several indicators based on orders waiting within order books and show that these indicators give agents relevant information that improve their trading behaviour.

In macroscopic systems, the usual reasoning is to perform a technical analysis of historical prices to derive a future increase or decrease. This is typically the case strategies *chartist* that seek specific forms within price curves.

To show the contribution of order book information for agents reasoning, we propose in this paper to start with a reference LCS agent called PriceLCS, reasoning only on past prices, and improve it by adding agents called OrderLCS which rely on prices but also use market microstructure indicators. Then, we show through a set of experiments that these indicators provide relevant and useful information to agents to improve their trading performances.

The PriceLCS only uses technical analysis indicators presented in Table 6 that can take into account price change on both short and long term on three types of simple criteria: price evolution compared to previous price (indicator 1), the average compared to previous price n (indicators 2–4), or from the middle of the range of values of n previous prices (indicators 5 and 6).

We now want to compare PriceLCS with better informed agents, OrderLCS that benefit not only from prices but also from pending orders.

Indeed, order books contain a lot of information, in particular, the gap between the best bid and best ask, called *bid-ask spread* (see Table 3). This

Table 6. Technical analysis indicators shared by `PriceLCS` and `OrderLCS`

Id	Market indicator definition
1	$p_t > p_{t-1}$
2	$p_t > 1/5 \times \sum_{i=t-1}^{t-5} p_i$
3	$p_t > 1/10 \times \sum_{i=t-1}^{t-10} p_i$
4	$p_t > 1/100 \times \sum_{i=t-1}^{t-100} p_i$
5	$p_t > 1/2[Minp_i + Maxp_i]_{i \in [t-1,t-10]}$
6	$p_t > 1/2[Minp_i + Maxp_i]_{i \in [t-1,t-100]}$

Table 7. Technical market indicators based on bid-ask spread

Id	Market indicator definition
7	$bestP_{Ask} - bestP_{Bid} < k_7$
8	$r_t < k_8$

value can be related to an asymmetry of information, or a uncertainty about the value of the security. Moreover, it is a common market liquidity measure, the higher is the spread, the higher is the risk for an agent to sell or buy an asset in a short time frame.

To effectively leverage information from order books, we propose to add new indicators to `OrderLCS` agents: those concerning the value and evolution of the *bid-ask spread* and those related to an imbalance between supply and demand. The indicators proposed here are examples of criteria based on the orders that may be used by agents and are used in the evaluation Sect. 4.

Bid-ask spread. One can argue about the use of the *bid-ask spread* absolute value (indicator 7 in Table 7), but we propose to use instead the following ratio: $r = \frac{bestP_{Ask}}{bestP_{Bid}}$ (indicator 8).

One can for example compare the current value of this ratio to previous value (indicator 9), or the average of k_{10} previous values (indicator 10), or in the middle of the range of k_{11} previous values of r (indicator 11), to determine if the current *bid-ask spread* value is rather high or low (Table 8).

Table 8. Market indicators based on ration evolution $r = \frac{meilleur P_{Vente}}{meilleur P_{Achat}}$

Id	Market indicator definition
9	$r_t > r_{t-k_9}$
10	$r_t > 1/k_{10} \times \sum_{i=t-1}^{t-k_{10}} r_i$
11	$r_t > 1/2[Minr_i + Maxr_i]_{i \in [t-1,t-k_{11}]}$

Table 9. Market indicators based on relative size of bid and ask pending orders

Id	Market indicator definition
12	No bid orders pending
13	No ask orders pending
14	$\frac{nbAsk}{nbBid} > k_{14}$

Indicators Based on Imbalance Between Supply and Demand. The relative size in number of orders on both sides of the order book is also a useful information because it may reflect an imbalance between supply and demand (Table 9). This imbalance in a way or the other, may announce an upcoming change price that the agent could benefit. Indicators 12 and 13 check if there are buy and sell orders in the order book, and indicator 14 reasons on the ratio $\frac{nbAsk}{nbBid}$.

However, the relative size of the two parts of the order book is not the best way to assess the imbalance between supply and demand. In fact, if 10 assets are sold at the same price p_0, the offer is better than if only one asset is on sale for p_0 and 9 others have a higher price, yet $\frac{nbAsk}{nbBid}$ ratio remains the same. We should therefore also take into account the price differences in the calculation of supply and demand.

$$offer = \sum_{order \in Ask} \frac{Quantity(order)}{Price(order) - bidAskMid}$$

$$demand = \sum_{ordre \in Bid} \frac{Quantity(order)}{bidAskMid - Price(order)}$$

with:

$$bidAskMid = \frac{bestP_{Ask} + bestP_{Bid}}{2}$$

$bidAskMid$ is the average of the best ask price and the best bid price. By dividing the quantity of each order by the difference between its price and $bidAskMid$, taking into account the fact that some orders have a price limit too high or too low to constitute an interesting offer or demand. We are then interested in the ratio $q = \frac{offer}{demand}$ (indicator 15) and its evolution (indicators 16–18).

OrderLCS agents use indicators 1–6 as PriceLCS agents, but they also use indicators based on orders (indicators 7–18). These order based indicators are only examples and many other indicators could be relevant. Nevertheless, the proposed indicators cover many aspects of order books and are widely accepted in finance. Indicators using a constant k have been implemented in several versions with several values of this constant.

Table 10. Market indicators based on imbalance within bid ask pending orders

Id	Market indicator definition
15	$q_t > k_{15}$
16	$q_t > q_{t-k_{16}}$
17	$q_t > 1/k_{17} \times \sum_{i=t-1}^{t-k_{17}} q_i$
18	$q_t > 1/2[Min q_i + Max q_i]_{i \in [t-1, t-k_{18}]}$

3   Methodology

Our study has two goals: to show that price based LCS are effective and that order based LCS are more effective. To do this, we first compare `PriceLCS` to various *chartist* agents to check that their learning mechanism allow them to get better results (first stage on the horizontal axis of the Fig. 1). In a second step, we want to demonstrate that indicators based on orders can improve LCS agents. For this, we compare `PriceLCS` with several `OrderLCS` agents (second stage on the horizontal axis in Fig. 1).

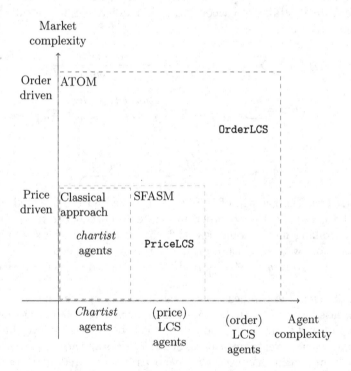

Fig. 1. Using a more complex market model allow the definition of better informed agents. We seek to demonstrate that LCS agents efficiency is better than chartists agents and then, that order based LCS agents outperforms price based LCS agents.

Evaluating and comparing agent behaviours is a difficult art. Firstly because, like voting systems, there are always methods that promote particular agents, and secondly because an agent is rarely good in absolute terms but often in relation to its opponents and the environment. Thus, we distinguish two problems: estimate an agent gain, that is to say its performance in a specific simulation, and to evaluate the agent, that is to say, estimating its performance being as general as possible.

3.1 Estimate Agents Performances

To compare agents, we must define a method to compute agents' gains, regardless of the type of evaluation selected. Two main criteria are possible here: either liquidity (*cash*) possessed by an agent or the richness of it (*wealth*), sum of its liquidity and the estimated portfolio value.

$$wealth = cash + \sum_{i=1}^{i<=assets} prix_i \times nbAssets_i$$

Although this estimation is questionable because it is approximative, it has the advantage of taking into account all the agent assets.

3.2 Evaluation Method

It is very difficult to say that an agent is better than another in absolute terms. In order to achieve the highest possible objective measures, it is important to test the agent in a sufficiently rich and representative selection to avoid potential biases due to a favorable or unfavorable environment. Moreover, the variation of the comparison set improves the objectivity of the measure. However, when an agent performs better than another in several sufficiently varied environments, then this agent demonstrated better robustness and we can assume that this agent is generally best.

Agent Comparison. Comparing agents by inserting them into a single environment is a major problem. It can exist specific relationship between two agents called "predator-prey" relationships, when these agents are in the presence of each other, the predator will maximize its results at the expense of prey. This, however, tells us nothing about the quality of each agent. If there is a relationship between two such agents that you want to evaluate, then the results for the two agents are distorted. This is why we use in our experiments a method that evaluates independently different agents to compare, based on the use of agents other than those to be evaluated. Finally, each type of agent should not be represented by a single individual, but by an agents population.

Populate Simulations. A good way to compare agents is to test their robustness to the variety of possible environments, and to average their performances. The performance obtained by an agent depends mainly on other competing agents, which is why it is necessary to vary types of agents encountered. To obtain diversified situations, simulations are populated with other agents types than those being evaluated. We call E all of these agents types, and S the set of agent types to evaluate. Some agents may be in S and E, if we try to evaluate them while they are also used to populate simulations. Agents that constitute E can be:

- *chartist* agents (moving average, RSI, momentum, variation, indicators and mixed moving average) using simple conditions on prices to predict price changes and decide whether to buy or to sell,
- periodic agents who buy and sell periodically
- *Zero Intelligence Traders* (ZIT, [13]) that send orders in a random direction with a random limit price,
- LCS agents based solely on price. These agents use simple market indicators (all of the form $p_t > p_{tX}$). These agents perform quite badly, but it seems important to us to have a minimum of adaptive agents within E.

Agent Families. To evaluate a trading behaviour, it is preferable to perform experiments with multiple instances of each agent instead of only one. In a real environment, it is rare that an individual is the only one to use a given strategy. So the concept of agent family, which is a set of agents using the same strategy and the same parameters has to be introduced. The use of agents families and the scaling that it induces has several consequences, including a better temporal distribution of the round of talk, possible interactions between multiple agents of the same type, or the smoothing of result for non deterministic agents.

3.3   Evaluation Methods

Once the measurement of individual performance is fixed, there are many ways to evaluate the performance of an agent in a community of n agents. Two main types of evaluation coexist [3]: evaluations in which n agents are placed in the same environment and compete with each other, and evaluations in which n agents are classified with respect to the same set of opponents. For each of these two types of evaluation, it is still possible to evaluate agents individually or with agents families (agents with the same type and parameters).

Various performance measures are then possible:

- n agents are evaluated in the same simulation. Agents are classified according to their exact gains at the end of the simulation.
- n agents families are evaluated in the same simulation. Agents are ranked on the average earnings of their families at the end of the simulation. Using agents families, results variability due to agents stochasticity is attenuated.

- each agent family is evaluated separately from the other by running it against a common reference population. The n agents are then classified on the average earnings of their families. Separating families that are compared avoids introducing a bias in the evaluation due to the dominance of one agent over another.
- agent families are evaluated against a common reference population within an ecological competition. At the beginning, all families have the same number of instances but at the end of the simulation, instances are adjusted according to their family score.

These different simulations are ranked on their pertinence, indeed each protocol is superseded by its followers. For example, the use of reference families avoids distorting the comparison between families who would have dominance relationships between them even if one is not better than the other. In this article, we have chosen the most complete of them, the ecological competition, because it introduces a variation of population that avoids mutual support between families of agents and thus ensures a better robustness of the results.

3.4   Ecological Competition

Ecological competition is a selection method inspired by biology and the natural selection phenomenon [16,18]. In such context, several families evolve in the same environment, such as animal or plant species sharing the same environment. As in nature, their populations vary over time, in such a way that the best families see their populations increase, while the others decline.

In our case, a competition is a series of simulations on the market. Each family begins with an identical number of individuals, the total population of the competition remains constant throughout the competition. After each generation, the population of each family is revalued based on its gain. After each generation a proportionality rule on the score of each family is applied to keep the total population constant.

A family score is the total gain of its instances during the simulation. But, in ATOM, as in real markets, *traders* have the opportunity to borrow money to purchase assets. The agent *cash* can then be negative as well as its *wealth*. It is therefore possible that a family has a negative total gain, which is a bit problematic to apply a proportionality rule. To solve this problem, we propose to subtract the gain of the worst agent p (gain of p are most often negative, then it will be an addition) to the gain of all agents a.

$$gain_a \geq gain_p \Rightarrow gain_a - gain_p \geq 0$$

In this way, the gain of p agent is reduced to 0 and that of all others is positive (as the total gain of each family). The total gain of a family f is the sum of the modified wealth of its agents. It is then possible to apply a proportionality rule for calculating populations of the various agent families.

The population volume of a given agent family at the end of an ecological competition represents its adaptation to a particular environment (other families competing agents) and its effectiveness in this environment.

It may seem surprising at first that ecological competition is preferred to a more classic economic competition, where an agent leaves the competition when it is ruined. However, this approach presents several problems:

- with the ability to borrow, it is difficult to determine whether an agent is ruined, this choice is arbitrary.
- an agent can decide to leave the market before going broke on it.
- this approach does not allow the opportunity for new agents to enter the market by playing a strategy they observed success.

Ecological competition has the advantage of presenting a dynamic environment that adapts itself to agents performances. This allows to experience a robustness against evolving environments.

3.5   Experimental Protocol

When an agent performs better than another in several various ecological competitions, we can assume that this agent is generally best. This is the method we used to evaluate our agents.

For each agent type to be evaluated $s \in S$, the following procedure is performed. 500 ecological competitions populated with one agent family of type s and various families of agents (5–20) from E (which contains currently forty agents), randomly selected each competition. 50 generations competition are generally sufficient to reach stable populations volumes in most cases. Each generation is composed of a trading day made of 2000 decisions per agent, which is necessary for learning agents to adapt themselves to their environment.

The average population of each family obtained after 500 competitions is a good measure of the performance of an agent, because it takes into account the robustness of the variety of possible environments, and the robustness to adaptive environments.

The first experiments consists in assessing the effectiveness of price bases LCS agents before comparing them to order based LCS agents.

3.6   Price and Quantity Policies Evaluation

The first experiment is to determine which pricing and quantity policy should be used with our LCS agents. To this aim, we compare a price based LCS with all policies variation against a reference population. As shown in Fig. 2, this experiment shows that the policy giving the best results is the policy that sends orders placed at the top of the order book, with a constant quantity. Thus, these policies are used throughout all other experiments.

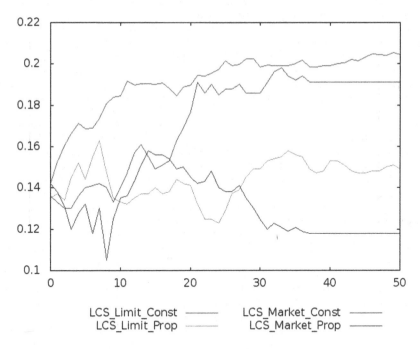

Fig. 2. Average PriceLCS family proportion with respect to the price and quantity fixing mechanisms: *Limit* to be at the top of the order book, *Market* for an immdiate execution, *Const* for a constant quantity and *Prop* for a proportional quantity.

4   Simulations Results

4.1   Price Based LCS Agents

The second experiment assesses the quality of price based learning. Figure 3 represents the average proportion of the population of various agents in an ecological competition. One of these families is `PriceLCS` and other agents are *simple* chartists (moving average, RSI, momentum ...) or agents with basic behaviours (periodic ZIT). It should be noted that price based LCS agents outperforms other families. Thus, this experiments demonstrates the effectiveness of the learning process achieved on prices analysis.

4.2   Order Based LCS Agents

The third experiment check that order based LCS agents perform better than LCS agents based solely on prices. Figure 4 represents the average population of a few agent families in an ecological competition. The `PriceLCS` uses indicators based on prices described in Table 6. For others families, these same price based indicators are used, but some order based indicators described in Tables 9 and 10 are added. 298 different agent types based on these additional descriptors have been tested, the Fig. 4 gathers only those that have obtain the best results.

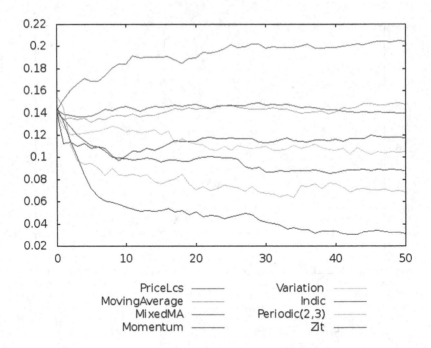

Fig. 3. Average population volume for each family: the `PriceLCS` graph is for price based LCS agents

It was found that the best families use 1–3 indicators taken from instances of indicators 10, 11, 16 and 17 (see Table 11). Other indicators have little influence or none at all on results. Indicators used by each family are described in Table 11. Each bit corresponds to a line. For example, agents of the family `OrderLcs_10001` use indicator 10 with $k = 100$ (first line table) and the indicator 17 with $k = 100$ (fifth row of the table).

We also observe that many of these families do better than the base agent. Agents using these good indicators have an average population at the end of the competition of about 100 % that of the base agent. This experiments demonstrates that LCS agents can be improved by the addition of order based indicators.

4.3   Assessing an Indicator Utility

We propose an utility measure u_i of an indicator i as the average score of rules for which the *trit* question is not undetermined (i.e. #). This score is calculated on all the rules of agents of a family at the end of a simulation.

If more indicators are used, the average utility of an indicator is small. To compare the usefulness of two indicators, we use the following ratio:

$$s_{ind} = \frac{u_{ind} \times nbIndicateurs}{\displaystyle\sum_{i \in indicateurs} u_i}$$

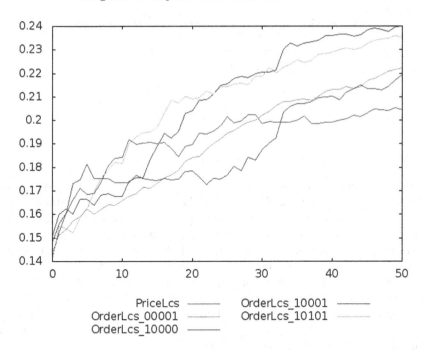

PriceLcs ———   OrderLcs_10001 ———
OrderLcs_00001 ———   OrderLcs_10101 ·········
OrderLcs_10000 ———

Fig. 4. Average volume population in an ecological competition: the `OrderLCS` graph is for order based LCS agents.

Table 11. Order based indicators that improve `PriceLCS` agents

Id.	k	Indicator instantiation
10	100	$r_t > 1/100 \times \sum_{i=t-1}^{t-100} r_i$
11	100	$r_t > 1/2[Minr_i + Maxr_i]_{i \in [t-1, t-100]}$
16	5	$q_t > q_{t-5}$
17	10	$q_t > 1/10 \times \sum_{i=t-1}^{t-10} q_i$
17	100	$q_t > 1/100 \times \sum_{i=t-1}^{t-100} q_i$

If $s_{ind} > 1$, then u_{ind} is higher than the average useful indicators of the observed LCS. Otherwise, if s_{ind} is less than 1, it means that u_{ind} is below the average. Therefore an indicator of quality maximizes the ratio s. We note on the Table 12 that this ratio is less than 1 for an indicator randomly returning true or false (line 1), less than 1 for indicators based solely on prices (line 2), but greater than 1 for several indicators based on orders (line 3), in particular those listed Table 11.

Table 12. Ratio value s for several kind of indicators

	Market indicator definition	s
1	Random	0.860
2	$p_t > 1/5 \times \sum_{i=t-1}^{t-5} p_i$	0.961
3	$r_t > 1/100 \times \sum_{i=t-1}^{t-100} r_i$	1.034

4.4 Learning Mechanism Saturation

The more information an agent has, the greater is its potential to achieve good prediction on price evolution. Nevertheless, the addition of an indicator expands the search space, and makes the learning process slower and sometimes less effective. All information are not relevant or pertinent. For example, one can think that $p_t > p_{t-1}$ is more relevant than $p_t - 100 > 101 - p_t$ for many situations.

Indeed, the expansion of the search space is the same regardless of the indicator added, so the information provided by an indicator must be significant enough to compensate for this expansion. In addition, if two indicators provide similar information, the added value of the second is low. We found that our LCS agents must therefore have a limited number of indicators (between 6 and 9), sufficiently differentiated from each other.

5 Conclusion

To achieve realistic financial simulations, it is important to populate artificial markets with adaptive agents. It allows to obtain price curves or price histories that are more realistic and consistent with stylized facts found in computational finance. It also enhances the experience of human agent who participates in a simulation with agents more *intelligent* and robust. It finally allows you to test the rules of market regulation in richer and more lifelike environments.

However, until now, the lack of software platforms both running orders driven markets and implementing a multi-agent approach, this type of simulation was done only on the price curve generated through equational models, as done by B. Arthur with SF-ASM. The ATOM platform, by its fidelity to order driven markets like *EuroNEXT-NYSE*, helps to push learning agents much further.

Learning in the context of a agent based artificial stock market is a complex process. It is often believed to one can learn from the behaviour of other agents. Recognizing other agent behaviour is a difficult thing, trying to adapt itself to an evolving environment is even more complex. It is also often believed that there is nothing to learn from stochastic agents and that only agents with deterministic behavior can bring more information. These suppositions do not take into account interesting information that are brought by the microstructure of double order books who sets prices in order driven markets. They plays a role similar to an accumulator. These are not necessarily the last orders to arrive that will be executed first, but the best deals. It is then possible make effective use of the information contained in these order books and have an *edge* on price trends. From these information can be found the best bid or best offer for sale of course, but also the size of the *bid-ask spread* or available quantity available, weighted or not by its distance.

In this article, after detailing the various possibilities to reason about orders and their consequences, we have shown how to set up learning classifier systems that take into account the information present in the market microstructure, that is to say pending orders in order books. To compare these agents, we have implemented an original adaptation of the principle of ecological competition that allows us to measure an agent performance and also its robustness to environmental changing. We were able to show that an agent who studies pending orders in order books is far more efficient than agents *chartist* or as his counterpart that reason solely on prices.

Further work is certainly needed in this direction, by varying the learning methods used and also by introducing indicators that attempt to recognize typical trading behaviours by studying orders evolution from a specific agents within order books. This, we consider this work as a first step towards more complete adaptive learning behaviours for artificial agents.

References

1. Arthur, W.B., Holland, J., LeBaron, B., Palmer, R., Tayler, P.: The Economy as an Evolving Complex System II, pp. 15–44. Addison-Wesley, Reading (1997)
2. Barbosa, R.P., Belo, O.: An agent task force for stock trading. In: Demazeau, Y., Pěchouček, M., Corchado, J.M., Pérez, J.B. (eds.) Adv. on Prac. Appl. of Agents and Mult. Sys., AISC, vol. 88, pp. 287–297. Springer, Heidelberg (2011)
3. Beaufils, B., Mathieu, P.: Cheating is not playing: methodological issues of computational game theory. In: ECAI'06 (2006)
4. Belter, K.: Supply and information content of order book depth: the case of displayed and hidden depth (2007)
5. Boland, E., Klingebiel, K., Stodgell, T.: The xcs classifier system in a financial market (2005)
6. Booker, L.B., Golbergand, D.E., Holland, J.I.: Classifier systems and genetic algorithms. Artif. Intell. **40**, 235–282 (1989)
7. Brandouy, O., Mathieu, P.: Efficient monitoring of financial orders with agent-based technologies. In: Demazeau, Y., Pěchouček, M., Corchado, J.M., Pérez, J.B. (eds.) Adv. on Prac. Appl. of Agents and Mult. Sys., AISC, vol. 88, pp. 277–286. Springer, Heidelberg (2011)
8. Mathieu, P., Brandouy, O.: A generic architecture for realistic simulations of complex financial dynamics. In: Demazeau, Y., Dignum, F., Corchado, J.M., Pérez, J.B. (eds.) Advances in PAAMS. AISC, vol. 70, pp. 185–197. Springer, Heidelberg (2010)
9. Brenner, T.: Chapter 18 Agent Learning Representation: Advice on Modelling Economic Learning. Handbook of Computational Economics, vol. 2, pp. 895–947. Elsevier, Amsterdam (2006)
10. Cao, C., Hansch, O., Wang, X.: The informational content of an open limit order book. In: EFA 2004 Maastricht Meetings

11. Cao, L., Tay, F.: Application of support vector machines in financial time series forecasting. Omega: Int. J. Manage. Sci. **29**, 309–317 (2001)
12. Cornuéjols, A., Miclet, L., Kodratoff, Y.: Apprentissage Artificiel Concepts et Algorithmes. Eyrolles, Paris (2002)
13. Gode, D.K., Sunder, S.: Allocative efficiency of markets with zero-intelligence traders: market as a partial substitute for individual rationality. J. Polit. Econ. **101**, 119–137 (1993)
14. Kozhan, R., Salmon, M.: The information content of a limit order book: the case of an fx market (2010)
15. LeBaron, B.: Building the santa fe artificial stock market. Brandeis University (2002)
16. Lotka, A.J.: Elements of Physical Biology. Williams and Wilkins, Baltimore (1925)
17. Mathieu, P., Secq, Y.: Environment updating and agent scheduling policies in agent-based simulators. In: ICAART'2012 (2012)
18. Volterra, V.: Variations and fluctuations of the number of individuals in animal species living together. In: Chapman, R.N. (ed.) Animal Ecology. McGraw-Hill, New York (1926)
19. Wilson, S.: Classifier fitness based on accuracy. Evol. Comput. **3**, 149–175 (1995)
20. Gershoff, M., Schulenburg, S.: Collective behavior based hierarchical XCS. In: Proceedings of the 9th Annual Conference Companion on Genetic and Evolutionary Computation, GECCO '07, London, UK, pp. 2695–2700. ACM, New York (2007)

Coupled K-Nearest Centroid Classification for Non-iid Data

Mu Li[1]($\boxtimes$), Jinjiu Li[1], Yuming Ou[1], Ya Zhang[2], Dan Luo[1], Maninder Bahtia[3],
and Longbing Cao[1]

[1] Advanced Analytics Institute, University of Technology Sydney, Ultimo, Australia
mu.li@student.uts.edu.au,
{jinjiu.li,yuming.ou,dan.luo,longbing.cao}@uts.edu.au
[2] Institute of Image Communication and Information Processing,
Shanghai Jiao Tong University, Shanghai, China
ya_zhang@sjtu.edu.cn
[3] Australian Tax Office, Sydney, Australia
maninder@ato.gov.au

Abstract. Most traditional classification methods assume the independence and identical distribution (iid) of objects, attributes and values. However, real world data, such as multi-agent data and behavioral data, usually contains strong couplings among values, attributes and objects, which greatly challenges existing methods and tools. This work targets the coupling similarities from these three perspectives and designs a novel classification method that applies a weighted K-Nearest Centroid to obtain the coupled similarity for non-iid data. From value and attribute perspectives, coupled similarity serves as a metric for nominal objects, which consider not only intra-coupled similarity within an attribute but also inter-coupled similarity between attributes. From the object perspective, we propose a more effective method that measures the centroid object by connecting all related objects. Extensive experiments on UCI and student data sets reveal that the proposed method outperforms classical methods for higher accuracy, especially in imbalanced data.

1 Introduction

Most of the existing classification methods make an assumption on the independence among values, attributes and objects [3]. For example, SVM [12] tries to classify objects by converting all the nominal features into binary numerical features [11]. More precisely, an attribute with ν distinct categorical values can be converted into ν separate new features. Consequently, this kind of methods believes different values of each feature share equal discriminative power in classification. However, in real world data, such as multi-agent interactions and agent behaviors, the coupling relationships and heterogeneity among data features and values are ubiquitous. This greatly challenges existing theories and systems.

Some similarity metrics try to measure distance by geometric analogies which represent the relationship of data values. For example, the similarity between 10

© Springer-Verlag Berlin Heidelberg 2014
R. Kowalczyk et al. (Eds.): TCCI XV, LNCS 8670, pp. 89–100, 2014.
DOI: 10.1007/978-3-662-44750-5_5

Table 1. Student information table

ID	Country	Education	Economic	Risk
1	USA	TAFE	H	H
2	Australia	TAFE	H	H
3	China	HighSchool	M	L
4	China	Bachelor	M	L
5	Japan	Bachelor	L	L
6	Japan	HighSchool	L	L

and 12 is greater than that of 10 and 2. There are many similarity metrics which have been explored for numerical data, such as Euclidean and Minkowski distances [8]. By contrast, similarity measurement over nominal variables has received much less attention. In a supervised learning process, heterogeneous distances [16] and modified value distance matrix ($MVDM$) [6] describe the similarity between categorical values. In unlabeled data, only limited research [8], such as simple matching similarity (SMS, which only uses 0s and 1s to distinguish similarities between distinct and identical categorical values) and occurrence frequency [2], discusses the similarity between nominal values. Reference [8] defined a specific similarity measure between attribute values, by extracting the intensity of the relationship between two data objects, which resemble each other frequently, which could obtain a larger similarity between them. The following is an problem illustrating this problem.

We conduct an educational data mining [14] project, which analyzes student behaviors on campus and assess the risk of failing in the target courses. In Table 1, six student are divided into two classes, one is high risk and the other is low risk. As shown in the table, there are three nominal features: country, educational background and economic status. Overlap similarity [13] measure the similarity between the countries "*China*" and "*Japan*" to be 0. However, *China* and *Japan* share a similar culture and characteristics, therefore, students from these two countries may have similar characteristics. Another observation by is that the similarity between "*Bachelor*" and "*HighSchool*" is equal to that of "*TAFE*" and "*HighSchool*". Intuitively, the similarity of the former pair should be greater since they drop into the same class L. Therefore, if students come from TAFE, they have a higher risk of failing in some courses than high school students or those who have a bachelor degree.

This example shows that it is difficult to analyze the similarity between nominal variables. Furthermore, numeric distance can not capture the significant correlation among nominal values. It is worthwhile to design an effective and efficient method to measure similarity among nominal variables.

Thus, we judge the similarity of categorical values by considering data characteristics. Two values are similar if they appear in a data set with similar frequency [2], which reflects the intra-coupled similarity within one feature. For example, two countries are similar if they appear with the same frequency, such as "*USA*" with "*Australia*" and "*China*" with "*Japan*". However, the reality is that the former pair is more similar than the latter. To improve the accuracy of intra-coupled similarity, it is believed that the object co-occurrence probabilities of attribute values induced on other features are comparable [1]. The similarity between countries should also cater for dependencies on other features such as "Education" and "Economic" over all movie objects, namely, the inter-coupled similarity between attributes. The coupling relationships between values and between attributes contribute to a more comprehensive understanding of object similarity [4]. No work that systematically considers both intra-coupled and inter-coupled similarities has been reported in the literature. This fact leads to the incomplete description of categorical value similarities, and apart from this, the similarity analysis on dependency aggregation is usually very costly.

In this work, in order to make the method more adaptive to real world data, we use the coupled similarity criterion to add both intra-coupled similarity and inter-coupled similarity to measure the coupled relation between data features. Based on the coupled distance (*COS*) measure by considering both Intra-coupled and Inter-coupled Attribute Value Similarities (*IaAVS* and *IeAVS*), which capture the attribute value frequency distribution and feature dependency aggregation with a high learning accuracy and relatively low complexity, respectively. We compare accuracies and efficiencies between our method and some classic method, we then evaluate our proposed measure with an existing metric on a variety of benchmark categorical data sets including real world educational application data, in terms of classification qualities.

This paper is organized as follows. In Sect. 2, we briefly review definitions of the coupled similarity matric. Section 3 proposes a novel the coupled similarities base classification method, and the demonstrate the efficiency and effectiveness analysis is given in Sect. 4. Section 5 gives an application example to show the practicalities of the proposed method. Finally, we conclude this paper in Sect. 7.

2   Related Work

References [2,8] discussed the similarity between categorical attributes. Cost and Salzberg [6] proposed *MVDM* based on labels, while Wilson and Martinez [16] studied heterogeneous distances for instance based learning. Some data mining techniques for nominal data [1,2] existed. The most famous are the distance measure and its diverse variants such as Jaccard coefficients [8], which are all intuitively based on the principle that the similarity measure is 1 with identical values and is otherwise 0. More recently, attribute value frequency distribution has been considered for similarity measures [2]; neighborhood-based similarities [10] are explored to describe the object neighborhood by using an overlap measure. These are different from our proposed method, which directly reveals the similarity between a pair of objects.

Recently, a number of researchers have pointed out that the attribute value similarities are also dependent on their coupling relations [2, 4]. Das and Mannila presented the Iterated Contextual Distances algorithm, convinced that the similarities among features and objects are inter-dependent [7]. Ahmad and Dey [1] proposed a computing dissimilarity matric by considering the value's co-occurrence. While the dissimilarity criterion of the latter leads to high accuracy, the computation is usually very costly, which limits its application in large-scale problems.

3   Problem Definition

In this section, *Coupled Attribute Value Similarity (CAVS)* is used in terms of both intra-coupled and inter-coupled value similarities. When we consider the similarity between attribute values, "intra-coupled" indicates the involvement of attribute value occurrence frequencies within one feature, while "inter-coupled" means the interaction of other features with this attribute. For example, the coupled value similarity between B_1 and B_2 concerns both the intra-coupled relationship specified by the repeated times of values B_1 and B_2: 2 and 2, and the inter-coupled interaction triggered by the other two features (a_1 and a_3).

Suppose we have the *Intra-coupled Attribute Value Similarity (IaAVS)* measure $\delta_j^{Ia}(x, y)$ and *Inter-coupled Attribute Value Similarity (IeAVS)* measure $\delta_j^{Ie}(x, y)$ for feature a_j and $x, y \in V_j$, then *CAVS* $\delta_j^A(x, y)$ is naturally derived by simultaneously considering both of them.

Definition 3.1. *Given an information table S, the **Coupled Attribute Value Similarity (CAVS)** between attribute values x and y of feature a_j is:*

$$\delta_j^A(x, y) = \delta_j^{Ia}(x, y) \cdot \delta_j^{Ie}(x, y) \tag{3.1}$$

where δ_j^{Ia} and δ_j^{Ie} are IaAVS and IeAVS, respectively.

3.1   Intra-Coupled Interaction

According to [8], it is a fact that the discrepancy of attribute value occurrence times reflects the value similarity in terms of frequency distribution. Thus, when calculating attribute value similarity, we consider the relationship between attribute value frequencies on one feature, and intra-coupled similarity which we use later, is defined as follow:

Definition 3.2. *Given an information table S, the **Intra-coupled Attribute Value Similarity (IaAVS)** between attribute values x and y of feature a_j is:*

$$\delta_j^{Ia}(x, y) = \frac{|g_j(x)| \cdot |g_j(y)|}{|g_j(x)| + |g_j(y)| + |g_j(x)| \cdot |g_j(y)|}. \tag{3.2}$$

Hence, by taking into account the frequencies of categories, an effective measure (*IaAVS*) has been captured to characterize the value similarity in terms of occurrence times.

3.2 Inter-Coupled Interaction

In terms of *IaAVS*, we have considered the intra-coupled similarity, i.e., the interaction of attribute values within one feature a_j. This does not, however, involve the couplings between other features $a_k(k \neq j)$ and feature a_j when calculating attribute value similarity. Accordingly, we discuss this dependency aggregation, i.e., inter-coupled interaction.

Definition 3.3. *The **Inter-coupled Relative Similarity based on Power Set (IRSP)** between attribute values x and y of feature a_j based on another feature a_k is:*

$$\delta_{j|k}^P(x,y) = \min_{W \subseteq V_k} \{2 - P_{k|j}(W|x) - P_{k|j}(\overline{W}|y)\}, \qquad (3.3)$$

where $\overline{W} = V_k \backslash W$ is the complementary set of a set W under the complete set V_k.

According to the above discussion, we can naturally define the similarity between the jth attribute value pair (x,y) on top of these four optional measures by aggregating all the relative similarities on features other than attribute a_j.

Definition 3.4. *Given an information table S, the **Inter-coupled Attribute Value Similarity (IeAVS)** between attribute values x and y of feature a_j is:*

$$\delta_j^{Ie}(x,y) = \sum_{k=1,k\neq j}^{n} \alpha_k \delta_{j|k}(x,y), \qquad (3.4)$$

where α_k is the weight parameter for feature a_k, $\sum_{k=1}^{n} \alpha_k = 1$, $\alpha_k \in [0,1]$, and $\delta_{j|k}(x,y)$ is one of the inter-coupled relative similarity candidates.

3.3 Coupled Similarity

After specifying *IaAVS* and *IeAVS*, a coupled similarity between objects is built based on *CAVS*. Then, we consider the sum of all these *CAVS*s analogous to the construction of Manhattan dissimilarity [8]. Formally, we have:

Definition 3.5. *Given an information table S, the **Coupled Object Similarity (COS)** between objects u_{i_1} and u_{i_2}:*

$$COS(u_{i_1}, u_{i_2}) = \sum_{j=1}^{n} \delta_j^A(x_{i_1 j}, x_{i_2 j}), \qquad (3.5)$$

where δ_j^A is the CAVS measure defined in (3.1), $x_{i_1 j}$ and $x_{i_2 j}$ are the attribute values of feature a_j for objects u_{i_1} and u_{i_2} respectively, and $1 \leq i_1, i_2 \leq m$, $1 \leq j \leq n$.

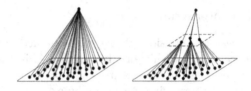

Fig. 1. Comparation times with and without clustering

4   Coupled Similarities Based Classification

In this section, we proposed a novel classification method based on the coupled similarity metric. Given a data set $\mathcal{D} = \{d_1, d_2, \ldots, d_n\}$, and $\Pi = \{\pi_1, \pi_2, \ldots, \pi_n\}$ are the classes of the data set. This work aims to extract more information between feature to feature and feature to class by applying coupled similarities, which can also been named coupled distance. In terms of the distance based classification task, the K-Nearest-Neighbors (KNN) is the most popular method; however, it lack of efficiency when it does the classification. In detail, when the KNN algorithm performs the classification task, it needs to compute the distance between the given object d_n to each of the other objects in $\mathcal{D}$ to find the K-nearest objects, and then to judge to which cluster this object belongs. In contrast, we proposed a method that only calculate the distance between the object to the cluster's centroid, which dramatically reduces comparison time Fig. 1, as the more training sets there are, the more time will be saved. This is a generalized process to find the most representative object to stand for the similar objects within one cluster. Moreover, this work also propose a novel method to improve the weaknesses of KNN classification. As KNN is based on equivalent significance to neighbors, this work adds weight to every object to enhance the discriminative power. The experiments show that the proposed method reduces of classification time substantially, without a loss of classification accuracy.

4.1   Clustering Within the Class with Coupled Similarities

In this section, a coupled similarities based clustering method is illustrated. Firstly, by Definition 3.5, a coupled similarity between two objects $COS(d_i, d_j)$ can be calculated. After this, for the classification task, we compute the coupled similarities within one class first because we assume there might be more coupled relations within one class than between two classes. In order to enhance the speed of the clustering process, we enumerated all the object s within the data set $\mathcal{D} = \{d_1, d_2, \ldots, d_n\}$ in to a comparison table (Table 2) and then calculated the coupled similarities between each of them. Since our definition of similarity is a relative value, it only can be applied when given two objects, which means it cannot create a middle point of two objects. Furthermore, the mean of two categorical attribute cannot be calculated as well, for instance, it is hard to say what gender is between male and female. As a result, a traditional clustering method like K-Means [15] cannot be applied directly, because it cannot find

Table 2. Coupled similarity between objects

Object pairs	Similarity
d_1, d_2	0.23
d_1, d_3	0.31
.	.
.	.
.	.
d_n, d_m	s

the mean point within a group of objects. To solve this problem, we used the Spherical K-Means [17] clustering method instead of K-Means as our clustering method.

Spherical K-Means Clustering using Coupled Similarities. Let d_1, d_2 be two categorical object from the data set $\mathcal{D} = \{d_1, d_2, \ldots, d_n\}$, the similarities among the objects is based on Definition 3.5. The clustering process partitions data set $\mathcal{D}$ into T clusters, and each of the clusters can be named as $\mathcal{C} = \{c_1, c_2, \ldots, c_t\}$ respectively. The perfect solution can be formally described as the following maximization problem:

$$\{c_t\}_{j=1}^k = \underset{\{c_t\}_{t=1}^k}{\arg\max} \sum_{t=1}^{k} \sum_{d_i \in c_t} Cos(m_t, d_i) \tag{4.1}$$

$\{c_t\} = \{d_{t1}, d_{t2}, \ldots, d_{tn}\}$ is a cluster with certain objects, A centroid point m_t of cluster c_t is an object within the c_t which has minimal similarity to all other objects within the cluster, for any object d' in c_t, the centroid point m_t that

$$\sum_{d_i \in c_t} Cos(m_t, d_i) \leq \sum_{d_i \in c_t} Cos(d', d_i) \tag{4.2}$$

The clustering method is straightforward, and is very similar to K-Means. Firstly, it randomly chooses K object from data set $\mathcal{D}$ as the centroid object m_k, m stands for the temp mode of the cluster and k is the cluster id for each cluster. Secondly, it allocates each object d_n to theirs nearest centroid object m_k as a intermediate cluster c_k, where c_k contains a set of objects $\{d_{k1}, d_{k2}, \ldots, d_{kn}\}$ which are the nearest objects to this centroid object m_k. Thirdly, it searches for a new centroid object within each cluster c_k, the new centroid object being the object which has minimal similarity to all other objects with in the cluster. When the new centroid object has been confirmed, the process is repeated to assign each object to the new centroid object to reform the cluster. Finally, it iterates the process until the centroid object is fixed for any cluster c_t.

$$m_t^n = m_t^{n+1} \tag{4.3}$$

n stands for the iteration times. Meanwhile, in some extreme case, the centroid object cannot be fixed at all, so there is an alternative criterion that

$$|(\sum_{t=1}^{k} \sum_{d_i \in c_t} Cos(m_t, d_i))^n - (\sum_{t=1}^{k} \sum_{d_i \in c_t} Cos(m_t, d_i))^{n+1}| \leq \varepsilon \qquad (4.4)$$

Similar to the above, n is the iteration time, ε is the certain threshold, if the "change" of the cluster after iteration is not significant, the searching algorithm will stop. The Spherical K-Means clustering method prevents the problems which K-Means leads to, thus it suitable for our coupled similarity based clustering.

Classification with Coupled Similarities Weighted Cluster Centroid. To simplify the problem, we take the binary classification task as an example. Once the clustering process has been finished, we have several clusters within both positive class π_A and negative class π_B. Moreover, each centroid of the cluster has its unique value for classification, due to the fact that difference of the coupled similarity between them is substantial. The coupled similarity cannot be expressed in 2D space, since all the similarities are relative and cannot be drawn on one flat picture. However, for simplicity, we use the following figure to illustrated the different similarities between clusters.

As Fig. 2 shows, each circular represents a cluster, and the caption of the circular C_{π_A} and C_{π_B} stands for the centroid object in both the positive class and negative class respectively. Moreover, we also use a different color to represent the clusters which belong to different classes. The ϕ_1 and ϕ_2 is the coupled similarity between two centroids $C_{\pi_{A3}}$ and $C_{\pi_{B5}}$ and $C_{\pi_{A2}}$ and $C_{\pi_{B2}}$. It is clear to see that, the coupled similarity ϕ_2 is significantly larger than ϕ_1. When undertaking a classification task, differences in the distance will affect the result significantly. Unfortunately, the classic KNN algorithm neglects this difference and judge every point as equivalently significant. More precisely, when it undertakes a classification task with an incoming object d_n, it only counts the amount of the k nearest neighbors $Count(\chi^k(d_n))$ which belong to each class, where the $\chi^k(d_n)$ denote the set of the nearest neighbors of object d_n, and if the number of neighbors belonging to class A is larger than the number of neighbors belonging to class

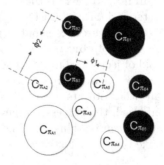

Fig. 2. The training data set after clustering

B, then it classify the object to class A, without considering the unique value of each of its neighbors. If F is the classification function, the classification process of traditional KNN can be describe as:

$$F(d_n) = \begin{cases} d_n \in \pi_A & Count(\chi^k(d_n) \in \pi_A) > \\ & Count(\chi^k(d_n) \in \pi_A) \\ d_n \in \pi_B & Count(\chi^k(d_n) \in \pi_B) > \\ & Count(\chi^k(d_n) \in \pi_A) \end{cases} \tag{4.5}$$

This work propose a novel classifier with a weighted cluster centroid, which comprehensively involves the information of the coupled similarity from every centroid object C_{π_A} in one class π_A to all other objects $d_{\pi_{\bar{A}}}^j$ belonging to the opposite class $\pi_{\bar{A}}$. The reason for this is that, by the concept of coupling, every object can be described by other objects, which have a relationship with it. This work utilizes this information to give centroid object, the computation of coupled similarities weight is quite straightforward:

$$W(C_{\pi_{An}}) = \sum_{i=1}^{m} \sum_{j=1}^{n} Cos(d_{\pi_A}^i, d_{\pi_{\bar{A}}}^j) \tag{4.6}$$

Finally, to classify an incoming object by accumulating the coupled similarities to every centroid object and adding these weights, the classification function becomes:

$$F(d_n) = \begin{cases} d_n \in \pi_A & \sum_{i=1}^{k} W(\chi^k(d_n) \in \pi_A) > \\ & \sum_{i=1}^{k} W(\chi^k(d_n) \in \pi_B) \\ d_n \in \pi_B & \sum_{i=1}^{k} W(\chi^k(d_n) \in \pi_B) > \\ & \sum_{i=1}^{k} W(\chi^k(d_n) \in \pi_A) \end{cases} \tag{4.7}$$

5 Experiment and Evaluation

In this section, empirical experiences on some UCI data will be explored. Without losing generality, we choose four UCI data sets for this experiment. More precisely, they are the sonar, hepatitis, horse-colic and SPECTF. In this experiment, we compared the proposed method to the two most classic classification method, the C4.5 decision tree and support vector machine (SVM). We not only compared their precision for the classification task, but also the ROC area of the classification result. The C4.5 decision tree set the confidence C to 0.5 and minimal count of leaf m to 2, the SVM set the cost c to 1 and eps e to 0.001, and our method set the cluster number K to one-tenth of the attribute's numbers. Since not all the features are categorical, if it is numerical we discrete it into 5 equal frequency categorical values. The C4.5 algorithm is used weka [9], and the SVM algorithm is used libsvm [5].

5.1   Classification Performance Comparison

The experiment results (Fig. 3) show that our proposed method "Coupled Distance based K-Nearest Weighted Centroid Classification" (CDK) outperform the classic method in all four data sets with different extensions according to the data distribution, which means, the performance of the proposed method is related to the data itself; that is, the more coupled relationship among the data, the more improvement of classification precision will be.

5.2   Classification Efficiency Comparison

Since not all the data sets have balance class distribution, precision is not the vital metric to evaluate the performance of the classifier. For instance, if a data set has 99 percent of data belonging to one class and the remainder belongs to another class, the classifier simply classifies all the data into the larger class, so it can reach 99 percent of precision for the classification task. This problem is serious in the SVM because it tries to minimize the miss-classified number. The experiment of the ROC area (Fig. 4) is aim to reveal this problem, the results show that the proposed method makes a significant improvement on the ROC area criterion, as the proposed method fully considers the class variation of the data in the weighted centroid classification process.

6   Application

In this section, we illustrate the performance of this novel method by applying it on a real world educational data set. Educational data mining is a growing hot research topic. A large number of studies mention the prediction of the risk of the probability of students failing their subjects. In this work, we used de-personalized students' demographic data on 400 students, with 80 demographic features in this experimental data. The demographic features include

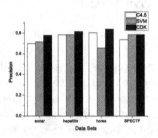

Fig. 3. Classification precision comparison

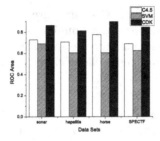

Fig. 4. Classification ROC area comparison

Table 3. Comparison on student data

Method	Precision	Recall	F-Measure	ROC area
C4.5	0.851	0.845	0.834	0.768
SVM	0.870	0.873	0.87	0.836
This work	**0.901**	**0.885**	**0.876**	**0.946**

their nationality, previous educational background, previous academic grades, previous scholarship records and more. Moreover, the data was labeled by students previous examination results, that is, if the student was in the top 30 percent of their peers, we labeled it as class A while others were labeled as B. The configuration of the experiment is the same as the aforementioned experiment. This experiment compared this novel method with other classic classification method by contrasting the standard classification task metric result, Table 3 shows that this work has an advantage over most of the metrics. More importantly, the result support our main concern about real world data, in that there are plenty of coupling relationships not only between the value of features, but also among the objects.

7   Conclusion

In this work, we have proposed a coupled distance based K-nearest weighted centroid classification method, and applied a coupled object similarity metric which involves both attribute value frequency distribution (intra-coupling) and feature dependency aggregation (inter-coupling) in measuring attribute value similarity for the classification of nominal data. Substantial experiments have shown that applied inter-coupled relative similarity measures significantly outperform the other method without considering the coupling relations. However, in terms of efficiency, in particular on large-scale data, to maintaining equal accuracy, our method does not show its advantages. Moreover, the dissimilarity metric is more comprehensive and accurate in capturing the clustering qualities in accordance with substantial empirical results and also has its superior result in the following classification tasks.

We are currently applying the *COS* measure and clustering based classification method for our educational data mining tasks. We are also considering extending the notion of "coupling" for the similarity of unstructured data such as text data. Moreover, the proposed classification method also has potential for other applications.

Acknowledgment. This work is sponsored by the Australian Research Council Grants (DP1096218, DP0988016, LP100200774, LP0989721), and Australian Research Council Linkage Grant (LP100200774).

References

1. Ahmad, A., Dey, L.: A k-mean clustering algorithm for mixed numeric and categorical data. Data Knowl. Eng. **63**, 503–527 (2007)
2. Boriah, S., Chandola, V., Kumar, V.: Similarity measures for categorical data: a comparative evaluation. In: Proceedings of the 8th SIAM International Conference on Data Mining, pp. 243–254 (2008)
3. Cao, L.: Non-iidness learning: an overview. Comput. J. 1–18 (2013)
4. Cao, L., Philip, S.Y.: Behavior Computing: Modeling, Analysis, Mining and Decision. Springer, Berlin (2012)
5. Chang, C.-C., Lin, C.-J.: LIBSVM: a library for support vector machines. ACM Trans. Intell. Syst. Technol. **2**, 27:1–27:27 (2011)
6. Cost, S., Salzberg, S.: A weighted nearest neighbor algorithm for learning with symbolic features. Mach. Learn. **10**(1), 57–78 (1993)
7. Das, G., Mannila, H.: Context-based similarity measures for categorical databases. In: Zighed, D.A., Komorowski, J., Żytkow, J.M. (eds.) PKDD 2000. LNCS (LNAI), vol. 1910, pp. 201–210. Springer, Heidelberg (2000)
8. Gan, G., Ma, C., Wu, J.: Data Clustering: Theory, Algorithms, and Applications. ASA-SIAM Series on Statistics and Applied Probability. SIAM, Philadelphia (2007)
9. Hall, M., Frank, E., Holmes, G., Pfahringer, B., Reutemann, P., Witten, I.H.: The weka data mining software: an update. ACM SIGKDD Explor. Newsl. **11**(1), 10–18 (2009)
10. Houle, M.E., Oria, V., Qasim, U.: Active caching for similarity queries based on shared-neighbor information. In: Proceedings of the 19th ACM International Conference on Information and Knowledge Management, pp. 669–678 (2010)
11. Hsu, C.-W., Chang, C.-C., Lin, C.-J., et al.: A practical guide to support vector classification (2003)
12. Joachims, T.: Making large scale svm learning practical (1999)
13. Li, C., Li, H.: A survey of distance metrics for nominal attributes. J. Softw. **5**(11), 1262–1269 (2010)
14. Romero, C., Ventura, S.: Educational data mining: a review of the state of the art. IEEE Trans. Syst. Man, Cybern. Part C: Appl. Rev. **40**(6), 601–618 (2010)
15. Teknomo, K.: K-means clustering tutorial. Medicine **100**(4), 3 (2006)
16. Wilson, D.R., Martinez, T.R.: Improved heterogeneous distance functions. J. Artif. Intell. Res. **6**, 1–34 (1997)
17. Zhong, S.: Efficient online spherical k-means clustering. In: Proceedings of the 2005 IEEE International Joint Conference on Neural Networks, IJCNN'05, vol. 5, pp. 3180–3185. IEEE (2005)

An Adaptative Multi-Agent System to Co-construct an Ontology from Texts with an Ontologist

Zied Sellami and Valérie Camps[(✉)]

IRIT, Université de Toulouse, Toulouse, France
{sellami,camps}@irit.fr

Abstract. Ontologies are one of the most used representations to model the domain knowledge. An ontology consists of a set of concepts connected by semantic relations. The construction and evolution of an ontology are complex and time-consuming tasks. This paper presents DYNAMO-MAS, an Adaptative Multi-Agent System (AMAS) that automates these tasks by co-constructing an ontology from texts with an ontologist. Terms and concepts of a given domain are agentified and they act, according to the AMAS approach, by solving the non cooperative situations they locally perceive at runtime. These agents cooperate to determine their position in the AMAS (that is the ontology) thanks to (*i*) lexical relations between terms, (*ii*) some adaptive mechanisms enabling addition, removing or moving of new terms, of concepts and of relations in the ontology as well as (*iii*) feedbacks from the ontologist about the propositions given by the AMAS. This paper focuses on the instantiation of the AMAS approach to this difficult problem. It presents the architecture of DYNAMO-MAS, and details the cooperative behaviors of the two types of agents we defined for ontology evolution. Finally evaluations made on three different ontologies are given in order to show the genericity of our solution.

1 Introduction

In the last ten years, ontology engineering from texts has emerged as a promising way to save time and to gain efficiency for the construction or the evolution of ontologies [10]. But texts do not cover all the required information to construct or evolve a relevant domain model, and human interpretation and validation are required at several stages in this process. That is why ontology engineering remains a particularly complex task [29].

Our contribution in this paper completes a previous work [38] that proposes an AMAS named DYNAMO-MAS[1] enabling to construct an ontology from texts. DYNAMO-MAS automatically proposes new concepts and/or terms to be evaluated by an ontologist. This paper presents the design and the evaluation of

---

[1] DYNAMO: DYNAMic Ontology for information retrieval; http://www.irit.fr/DYNAMO/; MAS: Multi-Agent System.

© Springer-Verlag Berlin Heidelberg 2014
R. Kowalczyk et al. (Eds.): TCCI XV, LNCS 8670, pp. 101–132, 2014.
DOI: 10.1007/978-3-662-44750-5_6

DYNAMO-MAS, an interactive software based on an AMAS that aims at evolving ontologies from text. Section 2 describes related works regarding existing tools for evolving ontologies from text. It also analyses the links between Multi-Agent Systems (MAS) and ontologies. Section 3 is devoted to the presentation of the AMAS approach that we used to implement DYNAMO-MAS. The overview of the DYNAMO project, the defined architecture as well as our approach for ontology evolution are detailed in Sect. 4. Section 5 expounds the cooperative behaviors of the two types of agents we defined. Section 6 contains the experiments of ontology evolution that were carried out with DYNAMO-MAS and an analysis of their results. We conclude and plan some future works in Sect. 7.

2   Related Works

This section gives a brief overview of systems dealing with automatic ontologies evolution from texts. A more general state of the art on ontologies evolution from texts can be seen in [37]. As this paper is devoted to the presentation of the core of DYNAMO-MAS, the related works section focuses on the links between ontologies and multi-agent systems.

2.1   Ontologies Evolution from Texts

Few existing works deal with the automatic evolution of ontologies from texts. Most of them focus on the construction of ontologies; if there is a need of updating the ontology, another ontology is built from scratch (Text Onto Miner [19], OntoLearn [45], Text-To-Onto [14] and DOGMA [33]). Some works concern the management of the ontology evolution process (usually manual) or the management and the comparison of different versions of an ontology [16,17,25,43]. Other works focus on the propagation of ontology modifications on some artifacts (other ontologies, applications, data,...) [25,43].

To our knowledge, only two systems propose automatic ontology evolution from texts. The first one, EVOLVA [49] uses results of terms extraction from texts as well as other ontologies to identify new concepts to include to the ontology. For each concept to add, it tries to retrieve if there is a relation between this concept and a concept already present in the current ontology. EVOLVA is only useful for evolving English ontologies. When the domain modeled by the ontology is very specific, EVOLVA has difficulties to detect relations between a new concept and current concepts of the ontology. A more detailed experiment and comparison between DYNAMO-MAS and EVOLVA is given in [50]. The second system is a first prototype of DYNAMO [31]. It is only able to construct an ontology from scratch but not to make it evolving. The agents of the MAS implement a distributed clustering algorithm that identifies clusters of terms from a large text corpus. These clusters lead to the definition of concepts as well as their organization into a hierarchy. Each agent represents a candidate term extracted from the corpus and estimates its similarity with others thanks to statistical features. Several evaluations conducted with this DYNAMO first prototype confirmed that statistical approaches [23] are inefficient when texts are short (it is our case in the context of our work).

Table 1 shows a comparison between DYNAMO-MAS and the previously presented tools for ontologies evolution.

- **Reconstruction** means that the tool evolves the ontology by the construction from scratch of a new one;
- **Incremental evolution** means that the tool evolves the current version of the ontology;
- **Tool availability** means that the tool is available, can be downloaded from the Web and easily used;
- **NLP Knowledge** means that the person who will use the tool needs to have knowledge in NLP;
- **Processed language** indicates the language of the corpus;
- **Depending on corpus size** means that the tool only works with large corpus;
- **For multiple domains** means that the tool is able to evolve ontologies coming from many domains.

Table 1. Comparaison of DYNAMO-MAS and other ontologies evolution tools.

Characteristics	Text Onto Miner	OntoLearn	Text-To-Onto	DOGMA	EVOLVA	DYNAMO First Prototype	DYNAMO MAS
Reconstruction	X	X	X	X		X	
Incremental evolution					X		X
Tool availability		X	X		X	X	X
NLP knowledge	Necessary	Not necessary	Not necessary	Not necessary	Not necessary	Not necessary	Not necessary
Processed language	EN	EN	EN	EN	EN	FR	EN-FR
Depending on corpus size	X	X	X	X		X	
For multiple domains	YES	YES	YES	YES	NO	YES	YES

Table 1 tells us that these tools are only effective for large corpus and are not suitable for small-volume corpus such as those of DYNAMO-MAS. Moreover, the user of TextOntoMiner has to be strongly qualified in NLP techniques to organize the processes of text analysis and ontology construction. Another difference is the attention given to the ontologist. In DYNAMO-MAS, we emphasize the notion of *interactive co-construction* of an ontology. This means the ontologist can accept, reject, or modify the proposals made by the system; the system has then to integrate the ontologist' s decisions to provide new proposals that the ontologist has then to check again, and so on. In other approaches, the tools are less interactive or not interactive at all (only the final ontology is proposed to the ontologist). This postpones the validation. Furthermore, these works manage in a different way ontology construction and ontology evolution, whereas we want to handle them uniformly. Finally, DYNAMO-MAS is a generic tool for ontology evolution from texts. A person who uses DYNAMO-MAS does not need to have NLP knowledge. DYNAMO-MAS can work with both small and large corpus in French or English languages.

2.2   Ontologies and Multi-agent Systems

The notion of ontology is often mentioned when talking about communication and interaction between agents. Indeed, ontologies are used to allow agents to communicate and interact. They provide a formal basis for modeling languages communications between agents. By sharing the same ontology, that is to say the same vocabulary, agents are able to understand the messages exchanged and to respond efficiently. Much of researches concerning communication between agents and MAS became interested in exploiting ontologies and reasoning about ontologies in agents. Today, the use of ontologies and the need to evolve these ontologies highlight new challenges especially in terms of supports and tools. Other works have then used MAS in systems for ontologies evolution.

Ontologies for Communication Between Agents. To communicate and interact appropriately, agents need to understand and share a common language. Ontologies have been developed in this direction, to provide formal vocabularies that depend on application domain.

Stable ontologies in agents. Several studies have integrated ontologies in MAS especially the Semantic Web and Web Services [18,21,26,44].

Jointly with MAS design, designers construct an ontology of the application domain (often with OWL[2]). This ontology then forms a knowledge support for agents and allows them to formulate messages and to reason about messages. Other systems include several ontologies in the MAS functioning.

Elmore et al. [15] uses a set of ontologies for the treatment of heterogeneous data in order to unify them. Specifically they offer a MAS which aligns a view of data from five laboratories. Each agent processes structured data of one laboratory in XML format. It also has an ontology describing the semantics of these data. All agents of the system have a common language of communication. Ontologies allow them to interpret the data sent in messages in order to construct a unified view of data.

Another system using ontologies is COMMA [8]. Each ontology is encapsulated in an agent and, in collaboration with an interface agent, helps the user to add (or to retrieve) documents in a knowledge base. For this, the implemented ontology agent is an aid in order to correctly construct the metadata of the new document or the query.

Evolving ontologies in agents. It is sometimes necessary to modify the ontology used by an agent or by a MAS when the application data evolve. Some works [2,26,40] propose to add mechanisms to agents (such as replacing a concept with another according to interactions with other agents) enabling them to modify an ontology. The aim of each agent is then to continue to communicate but not to construct a common view of a domain nor to make evolve an ontology. Each agent then has its own point of view on the application domain. Similarly, Gasser [47] and Viollet [46] propose ontologies alignment tools in order to ensure

---

[2] Web Ontology Language http://www.w3.org/2004/OWL/.

communications between agents when several ontologies are used. This enables to solve interoperability issues between heterogeneous agents.

Viollet uses ontologies to represent knowledge of agents [46]. To communicate, these agents exchange messages using FIPA-ACL. The semantic content of these messages is expressed using ontologies. Viollet adds then mechanisms of ontologies alignment to agents in order to relate different ontologies and therefore to understand the semantics contained in messages [46].

More recent works have focused on the learning of new concepts using the MAS paradigm. Afsharchi et al. [1] and Safari et al. [35] propose a MAS able to learn new concepts in order to answer to user's requests. The MAS is distributed over a set of sites (several universities). Each agent manages an ontology built from texts describing the proposed university programs (medicine, archeology, engineering, etc..). Each concept is described in extension, namely by the set of instances that describe it. The role of an agent is to answer to users' requests for university courses. To respond, the agent sometimes needs to learn a new concept because the query contains a new knowledge. For this, a learning agent (*learner agent*) sends a request to teacher agents (*teacher agent*) from other universities that contains the unrecognized concept or instance. The aim of (*teacher agents*) is to return the instances containing the words formed by the concept. Based on these results, the (*learner agent*) determines the largest intersection between the data of each (*teacher agent*) which corresponds to a new concept formed by instances that are shared with the others (*teacher agent*). However, the authors do not specify how the concept is inserted in the ontology.

Multi-Agent Systems for the Management of Ontologies. Few works use MAS for the construction or evolution of ontologies. MAS is often used as a tool to help the ontologist to seek knowledge or to check the consistency of an ontology. To our knowledge, only [30] used a MAS as an ontology.

Aldea et al. use a MAS within a platform for ontologies evolution [3]. This MAS consists of two types of agents: the *coordinator agent* and the *internet agent* whose roles consist in retrieving, weighting and ranking documents on Internet, in order to help a user to make his ontology evolve. The *coordinator agent* takes as input an ontology. This ontology is split into several parts. Each part contains one or more concepts. Then, the *coordinator agent* sends each part to the *internet agent* whose role is to retrieve pages that contain instances of these concepts. Finally, each *internet agent* returns its results in the form of sorted pages. These results are then presented to the user in order to help him to make his ontology evolve. The MAS role is not to make the ontology evolve but rather to bring documents that may contain new concepts or new instances.

Hadzic et al. provide a system for ontologies evolution using the agent paradigm [22]. Four types of agents have been defined: the *information agent*, the *data warehouse agent*, the *data mining agent* and the *ontology agent*. To make his ontology evolve, a user sends a request to the *information agent*. It is a request for collecting information about the modeled domain. The *information agent* role is to retrieve information on the current domain in databases. Then, it sends these data to the *data warehouse agent* to store them. Upon receipt of new data, the *data warehouse agent* asks the *data mining agent* to process data. The role of this latter is to extract new knowledge by applying *data mining* techniques to identify concepts as well as relations between concepts. Finally, it sends the result to the *ontology agent* that compares new knowledge with the current ontology. If differences exist, the *ontology agent* proposes to the user a list of changes. The user can then accept or reject the proposals. Mechanisms of this system are very similar to those used in [49]. The difference is that [49] do not use agents and rely on ontologies that are available on the Web rather than on databases. Nevertheless, [50] demonstrated that using external data sources becomes ineffective if the domain knowledge are very specific.

The construction and evolution of ontologies are a teamwork involving several people. Often these people work remotely on several versions of the same ontology. To manage this teamwork, several ontologies management collaborative tools using MAS have been proposed.

Bao and Honavar use agents in a collaborative tool for the construction of ontologies [7]. They propose only one type of agent called *interface agent*. It represents a user of the tool. Its role is to ensure the consistency of the modeled ontology according to users' concurrent modifications.

Slimani et al. propose a tool called $P^2OManager$ to manage the evolution of an ontology using a MAS [41,42]. The aim of this tool is to maintain the consistency between an ontology and the dependent ontologies using a set of agents. An agent can have three roles: *ontology agent*, *initiator ontology agent* and *dependent ontology agent*. The objective of these agents is to manage the changes of an ontology and their propagation to dependent ontologies. The role of the *ontology agent* is to detect if the ontology has changed or not. It is a "listener" that perceives the changes made on ontology by a user of the system. When this happens the agent changes its role and becomes *initiator ontology agent*. Its objective is then to detail all the changes that affected the ontology. Then the agent decides to propagate these changes to other agents having the *dependent ontology agent* role. For this, the *initiator ontology agent* checks for every change if there is a link between a modified element in the ontology and the dependent ontologies. If such a link exists, the agent sends a message to request the spread of change to the *dependent ontology agents*. Otherwise, the change is not propagated. Slimani et al. are interested in the propagation of ontologies changes to other dependent ontologies [41,42]. However, the automatic identification and extraction of new knowledge, as well as the automatic addition of new concept in the ontology are not available in this tool.

To summarize these works, an ontology usually enables communication between agents. To maintain communication when the application domain evolves or when heterogeneous agents are related, some works propose mechanisms of ontology alignments, of replacement of concepts, or of learning of new concepts or instances of concepts. The goal then is not to propose a domain ontology, but rather to enable agents to understand themselves. Other works use MAS in tools for evolution of ontologies. The aim of these MAS is to help a user to find new knowledge or to maintain the consistency of ontologies. The works proposed by [22,30] appear to us the most fully developed because they automate the extraction of knowledge and their integration in an ontology. However, for our problem of evolution of ontologies from texts (where texts are very short, with very specific knowledge and where no external knowledge-rich data sources exist) they seem difficult to be used.

3   Adaptive Multi-Agent Systems (AMAS)

The use of Multi-Agent Systems in order to evolve an ontology from texts seems to be a relevant idea. Indeed, a system for evolution of ontologies can be seen as a complex problem whose environment (additions of texts, ontologist's actions) is dynamic and opened. This system should be able to self-adapt to this environment. Several adaptation techniques exist but are ineffective for our particular problem because of constraints such as the small volume of data in our texts, the impossibility of defining the finality of the system, etc.. That is why an innovative adaptation using the concepts of emergence and self-organization seems to be relevant.

To solve the problem of ontologies evolution from texts, we need a system whose processings are distributed among several entities, each of which possesses a local view of its environment and is able to interact in an autonomous way to answer to a purpose. The dynamic environment of such a system and the complexity of our problem put in evidence the interest of using the MAS paradigm. The AMAS approach provides the theoretical foundations enabling the construction of such a system.

3.1   Functional Adequacy

The main asset of the AMAS approach [11,20] is to tackle the design of complex systems that can be incompletely specified and for which an *a priori* known algorithmic solution does not exist. It provides an organizational approach enabling the construction of multi-agent systems that continuously and locally self-adapt to the dynamics of their environment. It proposes to conceive an adaptive system while only focusing on the interactions between the system and its environment on the one hand and between the parts (agents) of the system on the other hand. These interactions are based on a local processing of the information by the components of the system that only have a local view of their environment. This principle of locality guarantees the emergent nature of the system functioning.

In this approach, the designer has only to define when and how each agent composing the system has to locally decide to change its interaction links with other agents in order to achieve the expected overall function (from the viewpoint of an external observer who knows its finality). In that case, the system is said "functionally adequate". We showed in previous works [11,20] that algorithms, which do not directly depend on the overall function to be obtained, are a solution for dynamically implementing systems able to self-adapt to their contexts. That is why local behaviors we propose to assign to agents do not depend directly on this expected overall function. Each agent, according to its local perception, the local rules it pursues and its local task to be achieved, can change or adjust its interactions with other agents of the system or the environment. The modification of the interactions between the parts of the system will lead to the transformation of the resulting overall function of the system. So, according to interactions between the multi-agent system and its environment, the organization between agents emerges and constitutes an answer to unforeseeable events.

3.2   When Does an Agent Need to Self-adapt?

To reach this functional adequacy, we proved [11, 20] that each autonomous agent, which follows a cycle composed of three steps (perception/decision/action), has to keep relations as "cooperative" as possible with its social (other agents) and its physical environment. The definition of cooperation we use is not conventional (resources sharing, common work, etc.); it is a social attitude to which an agent must comply. Our definition is based on three local meta-rules the designer has to instantiate according to the problem to be solved:

- Meta-rule 1 (C_{per}): Every signal perceived by an agent must be understood without ambiguity.
- Meta-rule 2 (C_{dec}): Information coming from its perceptions has to lead the agent to produce a new decision.
- Meta-rule 3 (C_{act}): This reasoning must lead the agent to make actions that have to be useful for other agents and the environment.

An agent, that simultaneously locally checks these three rules, is in a cooperative state. This means that this agent is situated at the best position in the current organization.

$$\text{Cooperation} = C_{per} \wedge C_{dec} \wedge C_{act}$$

On the contrary, an agent that does not locally check at least one of the three previous meta-rules, is facing a "Non Cooperative Situation" (NCS).

These cooperation failures can be assimilated to "exceptions" in traditional programming. Different generic NCSs were then highlighted: *incomprehension* or *ambiguity* if C_{per} is not checked, *incompetence* or *unproductiveness* if C_{dec} is not obeyed and finally *uselessness* or *competition* or *conflict* when C_{act} is not checked. These generic NCSs have to be instantiated according to the problems to be solved. This approach can be qualified as "proscriptive" because each agent in the system has, first of all, to anticipate, to avoid, and to repair the NCS that occur in its environment during the system functioning. Thus, the algorithm of a cooperative agent can be summarized by two main steps: (*i*) when an agent is facing a cooperative situation, it acts according to its partial function; (*ii*) when an agent is facing a NCS, it acts in order to come back to a cooperative state.

3.3   How Does an Agent Self-adapt?

This approach has important methodological implications: designing an AMAS consists in defining and assigning cooperation rules to agents. Concretely, the designer, according to the current problem to solve, has (i) to define the nominal behavior of an agent, then (ii) to deduce the NCSs to which the agent can be confronted with, and finally (iii) to define the actions the agent must perform to come back to a cooperative state and to self-adapt to the environmental dynamics. This self-adaptation of an agent is implemented by two main behaviors [9]. The **nominal behavior** directly related to the partial function of the agent that contributes to the overall emerging function; The **cooperative behavior** which includes the detection of and the resolution of NCS as well as the anticipation, the prevention of the occurrence of NCS. This behavior, responsible for specific adaptation process of the system, is subdivided into three behaviors:

1. The *tuning behavior* consists in analyzing the nominal behavior calculation in order to find cooperation failures. If there are, it tries to solve these NCS by modifying the parameters that take an active part in the nominal behavior.
2. The *reorganization behavior* consists in modifying the way in which an agent interacts with its environment and the other agents. This behavior is usually carried out when a uselessness or an incompetence NCS is detected.
3. The *evolution behavior* consists in creating new agents or in suppressing the current agent.

3.4   Methodological Impacts

This approach was applied successfully to the resolution of various types of problems related to different fields (user personalization [27], collective robotic [32], etc.). Obtained results encouraged us to promote the use of AMAS approach and to build a methodology named ADELFE[3] for designing adaptive systems. ADELFE only concerns applications in which self-organization makes the solution emerge from the interactions of their parts. It also indicates to the designer if the use of the AMAS approach is relevant for the construction of his application. If the AMAS approach is relevant, ADELFE helps him to express the behavior of the agents composing the system and the behavior of the society formed by these agents. ADELFE mainly focuses on the identification of all NCS that may appear during the system functioning and then on the definition of the actions the agents have to perform to come back to a cooperative state.

4   DYNAMO Overview

DYNAMO is an ANR[4] funded research project. Our contribution in this project was to propose a method and a tool that allow the construction and the evolution

---

[3] http://www.irit.fr/ADELFE
[4] http://www.agence-nationale-recherche.fr/

of Terminological and Ontological Resource (TOR) from a corpus of documents in order to facilitate semantic information retrieval. A TOR is a resource having a conceptual component (an ontology) and a lexical component (a terminology) [13,29]. A TOR contains not only a set of domain concepts but also a set of associated terms (their linguistic manifestations in documents: every term "denotes" at least one concept). These terms are used to annotate documents in order to do semantic information retrievals. This paper does not propose a new model for the representation of a TOR. Our TOR (called "ontology" in the rest of the paper) is formalized using the OWL-based TOR model and was provided by a partner of the DYNAMO project [34]. The TOR model used is a meta-model in which the OWL ontology concepts and associated terms are OWL classes (Fig. 1). A "concept" class is denoted by one or more classes "terms". Symmetrically, a "term" class must necessarily have a denotation link toward a "concept" class.

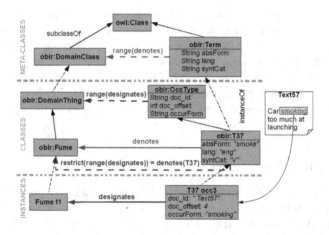

Fig. 1. The DYNAMO TOR model.

The core of our work was the definition and the conception of an AMAS called "DYNAMO-MAS" whose design principles follow the ADELFE methodology. Before detailing our system, we first present the main steps of our TOR evolution process from texts. Thereafter, we present the architecture of DYNAMO-MAS before explaining the cooperative behaviors of agents that we have defined.

4.1   Ontology Evolution Process

According to the principle of *ontological continuity* [48], we assume that the evolution of the domain knowledge does not affect the knowledge previously modeled in a TOR. Changing a TOR (ontology) consists then in adding other relations, other terms and/or other concepts. Our approach consists of 4 main steps (see Fig. 2). Initially, a TOR is modeled from a corpus of text documents.

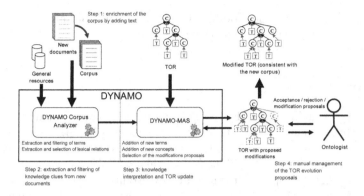

Fig. 2. The proposed process for the evolution of an ontology from texts.

This TOR is consistent with this corpus. The addition of new documents in the corpus triggers the process of TOR evolution.

Step 1: enrichment of the corpus by adding new text. This step consists in adding new documents to a corpus that will be then annotated by the TOR.

Step 2: extraction and filtering of knowledge clues from new documents of the corpus. This step consists in extracting (from new documents) candidate terms of the domain as well as lexical relations between these terms. This step also consists in identifying, among all the terms and lexical relations proposed by the NLP tools, those that are relevant to keep.

Step 3: knowledge interpretation and TOR update. This step consists in representing the terms and lexical relations (selected in the previous step) in the form of terms, concepts and relations between concepts, and in adding them to the current TOR. To do this, the system uses domain knowledge. Each new term to be proposed has to be connected (by a denotation link) to a concept (already existing or to be created) of the TOR. When a new concept is created, it has to be positioned in the TOR, and thus connected to one of the ontology concepts (if necessary to the TOP concept). To do this, the system uses lexical relations found in corpus, or uses the relations between concepts in a general ontology or in a general structured lexicon (WordNet or Wolf). When all the extracted terms are processed, some new concepts and new terms are selected by the system. The TOR, thus enriched by these most relevant new concepts and terms, is then proposed to the ontologist.

Step 4: manual management of the TOR evolution proposals. This step consists in finalizing the evolution of the TOR. To do this, the ontologist analyzes the TOR modifications proposals. He validates, modifies and/or rejects them one by one via the TOR editor. He can also "manually" add other concepts and other terms. He can also reorganize some parts of the ontology in order to better structure it. The system takes into account these modifications in order to update its internal representation of the TOR. This process ends when the ontologist is satisfied by the modified TOR and when, from his viewpoint, the TOR considered as "consistent" with the new corpus.

We propose to automate the second and the third steps of this process.

4.2   TOR Evolution Architecture

The DYNAMO architecture (Fig. 3) consists of (i) a corpus analyzer, (ii) an Adaptive Multi-Agent System (DYNAMO-MAS) and (iii) a graphical user interface (GUI). The input of DYNAMO is a corpus of documents. The output of DYNAMO is an OWL ontology. The goal of this paper is to focus on the instantiation of the AMAS approach to the ontology evolution issue. That is why only the main characteristics of the Corpus Analyzer and the GUI are given. More information about the corpus analyzer are available in [37].

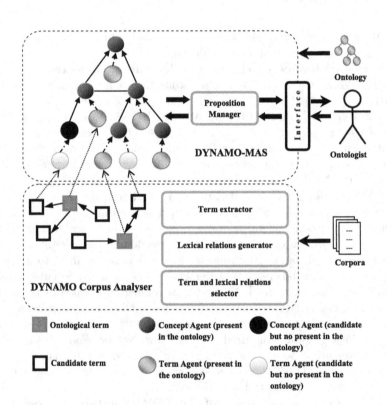

Fig. 3. DYNAMO architecture.

The goal of the **corpus analyzer** is to identify relevant candidate terms as well as relevant lexical relations that will be later agentified; it prepares the inputs for MAS. It includes a *terms extractor* named YaTeA [4] a *lexical relations generator* and a *term and lexical relations selector*. In this project, we are interested in four types of lexical relations: (*i*) *Hyperonymy* that expresses a generic-specific relation between terms; (*ii*) *Meronymy* that expresses a part-hood relation between terms; (*iii*) *Synonymy* that relates semantically close terms; (*iv*) Other relations (called *transverse relations*) that are any other kinds of lexical relations that will lead to a specific set of semantic relations, such as *causes, leads_to, etc.*

These lexical relations are extracted by three ways: (*i*) a lexico-syntactic patterns projection [24]; (*ii*) a syntactic dependency analysis between terms and candidate terms in order to extract hyperonym relations; and (*iii*) a similarity calculation between terms and candidate terms with the Levenshtein distance [28] to compute synonymy relations.

The *corpus analyzer* generates triplets $< T_i$, Rel, $T_j >$ where T_i and T_j are candidate terms or terms (whether the term belongs or not to the ontology) and *Rel* is a lexical relation. Each triplet has a confidence (Q, I) where Q is the quality of the relation (value between 1 and 10) and I is the number of instances of the relations in the corpus. The triplets are the inputs of the MAS.

The **MAS** receives as input, the triplets provided by the *corpus analyzer* and possibly an existing ontology. As output, it provides a modified ontology in the form of an OWL file respecting the used TOR model. DYNAMO-MAS also contains a component called *Nest* (not shown in Fig. 3) that is rather technical (it manages the creation of agents and their communications).

A **GUI** is implemented in the ontology editor Protégé[5]. It enables the ontologist to visualize the ontology as well as the MAS proposals. Through this interface, the ontologist can validate, delete or modify the DYNAMO-MAS proposals. He can also manually add other terms, concepts or relations. They will then be added to the MAS and they will lead to new proposals.

5   DYNAMO-MAS: An AMAS for TOR Evolution

DYNAMO-MAS consists of two components: (*i*) a MAS that represents the current TOR and (*ii*) a Proposition Manager whose role is to manage the MAS proposals as well as the interactions between the ontologist and the MAS.

We used ADELFE [36] to determine and define the two types of agents composing our MAS: (*i*) *term* agents that represent the terminological component of the TOR and (*ii*) *concept* agents that represent the conceptual part of the TOR.

The initial state of the MAS is an agentified TOR. The concepts of the TOR are *concept* agents connected by conceptual relations. The terms of the TOR are *term* agents connected to *concept* agents by denotation relations. The addition of a new text to the corpus triggers the corpus analyzer that identifies candidate terms and lexical relations between candidate terms and/or terms. The agentified candidate terms as well as lexical relations to be processed are then added to the MAS. When new *term* agents and new *concept* agents appear in the MAS,

---

[5] http://protege.stanford.edu/

they have to locally find their best position in the organization. It is the local goal of every agent. To achieve this goal, each agent has a nominal behavior and a cooperative behavior that subsumes the first one according to the AMAS approach (Subsect. 3.3).

5.1   Term Agent Behaviors

Term agents represent the terminological part of a TOR. A *term* agent has a status (*term* or *candidate term*) indicating whether the agent is part of the TOR (that is to say, a valid *term* agent) or is at the proposal stage (an invalid *term* agent). Each *term* agent is linked to other *term* agents in accordance with the lexical relations extracted from the corpus. It must also be linked to at least one *concept* agent according to the TOR model. Each relation between *term* agents is tagged by the confidence of the triplet $< T_i, \text{Rel}, T_j >$.

Term Agent Nominal Behavior. The goal of a *term* agent is to find its best position in the MAS and to propose itself to the ontologist. To do this, it must achieve three objectives: (*i*) to denote a *concept* agent ; (*ii*) to process all its outgoing lexical relations ; (*iii*) if the conditions are met, to propose itself to the ontologist. During its life cycle, each *term* agent processes its goals and the received requests (messages from other *term* agents or *concept* agents), from the highest priority to the lowest priority. A *Term* agent process its outgoing lexical relations from the most confident (most priority) to the less confident (less priority). Its priority objective is to denote a *concept* agent. Once this objective achieved, the next one is determined according to the confidence of the relation to be processed, the relevance of the agent for proposing itself to the ontologist and the confidences of the requests it receives. The algorithm explaining the nominal behavior of a *term* agent can be seen in [37].

To achieve its first objective, a *term* agent asks for the creation of a *concept* agent to the *Nest* tool. This creation is done if, in the current MAS, a *concept* agent having the same label does not exist. The *Nest* tool transmits thereafter the identifier of this new *concept* agent to the *term* agent. Then, the *term* agent sends to the *concept* agent a request for establishing a denotation relation (**❶**). This request is always accepted by the *concept* agent (**❷**). The confidence of the denotation relation is equal to the greatest confidence of the lexical relations of the *term* agent.

NCS1: At its creation a *term* agent may be faced with a *uselessness* NCS. Indeed, if a *term* agent is not linked to a *concept* agent, it cannot achieve its objectives; it is therefore useless.

NCS1 solution: To solve this NCS, the *term* agent sends a request to create a *concept* agent to the *Nest* tool. It sends then a denotation request to this *concept* agent. During its lifecycle, the *term* agent may change again its neighborhood or disappear if it becomes again useless.

To achieve its second objective, a *term* agent processes its outgoing lexical relations. A lexical relation has a confidence and a status (*not treated, treated or refused*). A *term* agent processes its relations from the most relevant (having the greatest confidence) to the less relevant. To do this, a *term* agent sends a request to its *concept* agent in order to transform the lexical relation ❸ (Fig. 4). The *concept* agent processes the request, then notifies the *term* agent with a message of acceptance or refusal ❹. The *term* agent updates the status of the processed relation (*treated or refused*). If the relation is refused, a *term* agent may later request to process the refused relation if its confidence increased. When a *term* agent asks for a synonym relation ❺ processing (Fig. 5), its *concept* agent sends a denotation request ❻ to the target *term* agent of this relation. If the confidence of the request is higher than the current denotation link of the target *term* agent, this latter accepts the request, changes its denotation relation ❼ and notifies the *concept* agent by a message of acceptation ❽. The target *term* agent refuses the request otherwise. The initial *term* agent is then notified ❾.

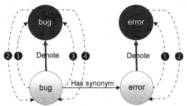

Fig. 4. Interactions between agents.

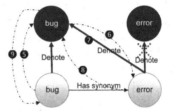

Fig. 5. Interactions between agents.

NCS2: During the processing of a denotion request sent by a *concept* agent, a *term* agent may be faced with an ambiguity NCS. The *term* agent cannot choose the *concept* agent that it wants to denote if the confidence of the request denotation $(Q_1; I_1)$ is equal to its current denotation confidence $(Q_2; I_2)$.

NCS2 solution: A *term* agent prefers a neighborhood having valid *concept* agents. In this way, to solve an ambiguity NCS, it chooses to denote the valid *concept* agent. When the *concept* agent that has sent the denotation request and the *concept* agent that is currently denoted by the *term* agent are both valid or invalid, the *term* agent chooses the agent having the greatest confidence (the most relevant according to it). In case of a tie, the *term* agent refuses the denotation request.

To achieve its third objective, a *term* agent calculates its relevance value (a score between 0 and 10). When this score is above a threshold (fixed to 5 but adjustable), a *term* agent proposes itself to be part of the ontology by sending a request to the Proposition Manager. The ontologist can tune this threshold starting with a low value at the beginning of the evolution process (to get a lot of suggestions) and increasing this value when the ontology provides good annotations for all the documents in the corpus. When the ontologist accepts or rejects a proposal, the Proposition Manager notifies the *term* agent with a rejection or an acceptance. The *term* agent updates then its status. To prevent *term* agents to propose themselves again after a rejection, a high value (equal to 9) is assigned to the ontologist's interventions (this parameter is adjustable). Indeed, during the ontology evolution, the ontologist can qualify a candidate term as irrelevant for the domain and can eventually later revise his decision. In this case, a *term* agent may propose itself again if its relevance value exceeds this reject value.

To summarize, a *term* agent tries to find its best position in the MAS organization by processing lexical relations. It moves from a *concept* agent towards another *concept* agent essentially by processing synonymy relations. To locally assess the adequacy of its position, the *term* agent calculates its relevance:

$$termAgentRelevance = \alpha_1 * P_1 + \alpha_2 * P_2 + \alpha_3 * P_3 + \alpha_4 * P_4$$

where P_1 is the maximum value of all its lexical relations; P_2 is the accuracy of its neighborhood; P_3 expresses the accuracy of the *term* agent's lexical relations; P_4 expresses the diversity of the *term* agent lexical relations and $\alpha_1, \alpha_2, \alpha_3, \alpha_4$ are the different weights of the P_i.

More precisely:

- P_1 = MaxLexicalRelationConfidence;
- P_2 = (nbTermAgentInTOR - nbTermAgentNotInTOR)/(nbTermAgentIn-TOR + nbTermAgentNotInTOR);
- P_3 = (nbAcceptedLexicalRelation - nbRefusedLexicalRelation)/(nbAccepted-LexicalRelation + nbRefusedLexicalRelation);
- P_4 = nbDifferenteLexicalRelation/nbAllDifferentLexicalRelation.

After various experiments with DYNAMO-MAS we empirically fixed the values of α_1 to 0.5, α_2 to 2, α_3 to 2 and α_4 to 1. These values best weighed the parameters of the agent relevance.

Term Agent Cooperative Behavior: The cooperative behavior of a *term* agent is defined through two cooperative behaviors (according to the AMAS approach): a reorganization behavior and an evolution behavior. The **reorganization behavior** occurs when a *term* agent denotes a valid *concept* agent (or when it receives from its *concept* agent a message telling it is a son of a valid *concept* agent).

For example (Fig. 6), `Main Frame`-term informs `Frame`-term that it denotes `Frame`-concept (❶). It also sends the relevance score of `Frame`-concept. This information is useful to a *term* agent that is the target of an hyperonymy relation. Indeed, in order to increase its confidence, `frame`-term may decide, when it receives this message, to move towards another *concept* agent. For this, it asks to `Main Frame`-concept which is its father *concept* agent (❷). `Main Frame`-concept then sends `Window` (❸). `Frame`-term is then faced with three denotation possibilities: either it can decide to not move and to remain linked to `Frame`-concept, or to denote `Main Frame`-concept, or to denote `Window`-concept. Its decision depends then on the following rule:

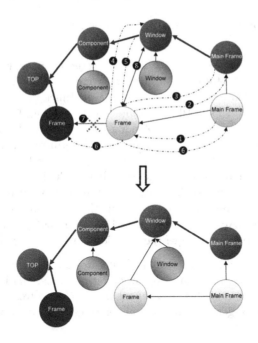

Fig. 6. Cooperative self-organization between *term* agents.

Rule: A *term* agent prefers denoting a valid *concept* agent than a invalid one because this can increase its relevance.

This rule enables to eliminate **Frame**-concept because it is an invalid *concept* agent. The *term* agent has then to choose between **Main Frame**-concept and **Window**-concept. The agent uses then the following rule:

Rule: A *term* agent prefers to denote a *concept* agent whose label is equal to its.

This more semantic rule was incorporated into the *term* agents by considering that a term is often mistaken with a concept. When this rule is not used by the *term* agent, it then uses the more abstract following rule:

Rule: In case of the reception of a denotation request from a *concept* agent, a *term* agent prefers to denote a *concept* agent whose relevance is the greatest (because the new denotation link will have a confidence equal to the confidence of the *concept* agent). The higher the confidence, more it maximizes the confidence of the *term* agent and thus enables it to propose itself to the ontologist. As this rule is based on confidences, an agent may be faced with a NCS.

NCS3: A *term* agent is facing an ambiguity NCS when it cannot choose between the *concept* agents towards whom moving, because their relevance are equal.

NCS3 solution: To solve this NCS, the *term* agent chooses to move towards **Window**-concept and not towards **Main Frame**-concept. This resolution is semantics. Indeed, a hyperonymy relation is often assimilated to an *is_a* relation.

In this sense, a *term* agent is more likely to be at a best position if it denotes Window-concept.

In Fig. 6, Frame-term decides to move to Window-concept whose relevance $(Q; I)$ is equal to its denotation relation with Frame-concept (❹). Window-concept accepts the denotation request and sends an acceptance message to Frame-term (❺). When it receives this notification and before moving, Frame-term informs its neighbors that it moves towards Window-concept (❻). This message enables agents to update their knowledge. Finally, Frame-term removes the old denotation relation (❼) and creates a new denotation relation with Window-concept (❽) with a confidence equal to the confidence it had with the former *concept* agent (because Frame-term agent is the initiator of the denotation relation request).

The **evolution behavior** occurs during the detection and elimination of useless *term* agents. According to the AMAS approach, all agents in a MAS must be useful in order to achieve the functional adequacy of the system.

NCS4: An intermediate *term* agent is facing a uselessness NCS when (i) it is linked to a *concept* agent denoted by other *term* agents; (ii) it is the target of hyperonymy relations and this *term* agent has the same frequency as the *term* agents whom it is target of hyperonomy relations; (iii) it is not a target of a hyperonymy relation.

NCS4 solution: When a *term* agent is facing a uselessness NCS, it disappears from the MAS. Before disappearing, a *term* agent informs its neighboring agents in order to allow them to update their knowledge (❶) (Fig. 7).

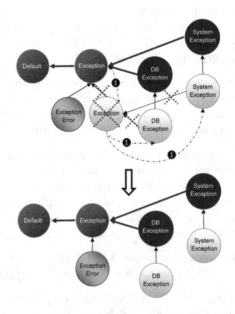

Fig. 7. Local treatment of a uselessness NCS by a *term* agent.

5.2   Concept Agent Behaviors

Concept agents represent the conceptual part of a TOR. A *concept* agent has a status (*concept* or *candidate concept*) indicating whether the agent is part of the ontology or is at proposal stage. A *concept* agent is linked to other *concept* agents by conceptual relations. It is also linked to *term* agents by denotation links. Every relation has a status (*not treated, treated or refused*) and a (Q, I) confidence.

Concept Agent Nominal Behavior: The goal of a *concept* agent is to find its *best* position in the MAS and to propose itself to the ontologist. To do this, it must achieve three objectives: (*i*) to have semantic relations with *concept* agents and denotation links with *term* agents; (*ii*) to determine a preferred label and (*iii*) to propose itself to the ontologist. Each *concept* agent processes its goals and the received requests from the highest priority to the lowest priority. The algorithm explaining the nominal behavior of a *concept* agent can be seen in [37].

To achieve its first objective, a *concept* agent has to consider requests from *term* agents ❶) to process lexical relations, and requests from *concept* agents (to establish conceptual relations (❷)). We assumed as in [5, 12, 39] that to each lexical relation will match a specific conceptual relation. A *hyperonymy* may lead to define an *is_a* relation between the concepts denoted by these terms; a *meronymy* may lead to define a *part_of* or an *ingredient_of* or a *member_of* relation between the concepts denoted by these terms; a *synonymy* may lead to connect the related terms to the same concept with a *denote* relation; *other relations* may lead to define specific semantic relations between the concepts denoted by the related terms, such as *causes, contributes_to, affects*, etc..

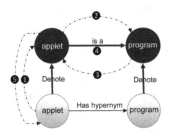

Fig. 8. Establishment of a *is_a* relation.

These principles of transformation are not generalizable to all results provided by the Corpus Analyzer because there may be extraction errors or conflicting lexical relations. A *concept* agent first operates on the basis of these principles, but it may afterwards change its relations and/or delete them. When a *concept* agent processes a request from a *term* agent to process a lexical relation, it sends a request to the concept agents denoted by the *term* agent that is target of the lexical relation (❷). For example (Fig. 8), a request to process a hyperonymy is sent to applet-concept. This message corresponds to a request to establish an *is_a* relation (❷). This request is sent by `applet`-concept towards `program`-concept agent (❷).

According to the AMAS approach, a cooperative agent must anticipate the appearance of NCSs that may affect the agent or its neighborhood. That is why, when a *concept* agent receives a request to process a lexical relation, it anticipates the appearance of conflicting relations:

1. A hyperonymy relation must not be processed if: (*i*) a *concept* agent is target of a *is_a* (resp. *has_part*) relation with the *concept* agent denoted by the *term* agent that is target of the hyperonymy relation and (*ii*) if that hyperonymy relation has a confidence $(Q_1; I_1)$ lower than the confidence $(Q_2; I_2)$ of the *is_a* (resp. *has_part*) relation. For example, `applet`-concept will not process the hyperonymy if it is the target of a *is_a* (resp. *has_part*) relation with `program`-concept that has a higher confidence.
2. A meronymy relation must not be processed if: (*i*) a *concept* agent is target of a *is_a* (resp. *has_part*) relation with the *concept* agent denoted by the *term* agent that is target of the meronymy relation and (*ii*) if that meronymy relation has a confidence $(Q_1; I_1)$ lower than the confidence $(Q_2; I_2)$ of the *is_a* (resp. *has_part*) relation. For example, `applet`-concept will not process the meronymy if it is the target of a *is_a* (resp. *has_part*) relation with `program`-concept that has a higher confidence.

A *concept* agent may have to choose between two conflicting relations in order to keep only one. When two conflicting relations have the same confidence (very rare), the agent detects a NCS.

NCS5: A *concept* agent is facing an ambiguity NCS concerning two conflicting relations R_1 and R_2 if the confidence of R_1 is equal to the confidence of R_2.

NCS5 solution: If among the relations R_1 and R_2 one and only one is a hyperonymy relation to be processed (resp. to be kept) or is an *is_a* relation, the *concept* agent will then prefer to process (resp. keep) this relation.

A *concept* agent prefers hierarchical relations (hyperonymy or *is_a*) compared to non-hierarchical relations (meronymy or *has_part*). This rule enables to solve the NCS. When this decision is semantically false, it will be checked again by the *concept* agent if the confidences of the lexical relations processed by the *term* agents that are targets of conceptual relations are reassessed when new texts are added to the corpus.

If a *concept* agent refuses a request to process a lexical relation, it sends a notification message to the *term* agent with a confidence equal to the confidence $(Q; I)$ of the conflicting relation. For example, in Fig. 8, `applet`-concept sends to `program`-concept a request for the establishment of an *is_a* conceptual relation (❷). This latter can accept or reject it by sending a notification to the agent (❸). A *concept* agent may refuse the establishment of a conceptual relation. Indeed, some combinations of conceptual relations are wrong and must not be established between *concept* agents. For this, a *concept* agent has the following knowledge: (*i*) an *is_a* relation is not symmetric; (*ii*) a *has_part* relation is not symmetric; (*iii*) an *is_a* relation between two *concept* agents cannot exist with a *has_part* relation between these two agents.

A request to establish a conceptual relation will be accepted only if this relation has a confidence $(Q; I)$ bigger than the confidence of the conflicting conceptual relation. In case of equal confidences, a NCS is detected.

NCS6: A *concept* agent A_1 is facing to an ambiguity NCS concerning two conceptual relations (it is the target of the first relation R_1 and the second relation, R_2, comes from a request for a conceptual relation establishment from a *concept* agent A_2), if both relations have equal confidences.

NCS6 solution: The A_1 agent compares its relevance P_1 with the relevance P_2 of the agent A_2. If P_2 is higher than P_1 then the A_1 agent accepts the establishment of the relation, otherwise it refuses. When these relevance are equal, the *concept* agent refuses the establishment of the required relation. This relation will be asked again if the lexical relations that led to its apparition in the AMAS evolve when new texts are added to the corpus.

When it receives a notification, the *concept* agent updates the status of the concerned relation as well as its relations with *concept* agents. In our example, *applet*-concept creates an *is_a* relation with *program*-concept (❹). As, in our AMAS, a *concept* cannot be polysemous, the establishment of an *is_a* relation leads to the displacement of a *concept* agent A of an old *concept* agent B towards another *concept* agent C. Finally, the *term* agent that is target of the lexical relation is notified by the *concept* agent that established the conceptual relation (❺). Finally, the processing of synonymy relations leads to the displacement of *term* agents from a *concept* agent towards another. In this case, it is possible that a *concept* agent has no denotation relation. This is a NCS, because without a *term* agent a *concept* agent cannot pursue its objectives.

NCS7: A *concept* agent is facing a uselessness NCS if it has no denotation relation.

NCS7 solution: The *concept* agent disappears from the MAS. Before disappearing, it informs the single *concept* agent to which it is connected with a *is_a* relation.

To achieve its second objective (to determine a preferred label), a *concept* agent chooses the label of the *term* agent having the denotation relation with the highest confidence. This label can evolve if new *term* agents denote the *concept* agent or if the confidences of the denotation links evolve.

To achieve its third objective (propose itself to the ontologist), a *concept* agent calculates its relevance value. When this value is above a threshold, a *concept* agent proposes itself to be part of the ontology by sending a request to the Proposition Manager.

An invalid *concept* agent considers itself as relevant if two criteria are checked:

1. The *concept* agent believes it is at the best position in the AMAS organization. This criterion is estimated by the parameters P_1, P_2, P_3, P_4;

2. the *concept* agent believes that it is an interesting *concept* agent to add to a TOR, that is to say it is closer to the "valid" status than the "Invalid" status. This criterion is estimated by the parameter P_5.

A *concept* agent processes its local relevance according to the following formula:

$$conceptAgentRelevance = \alpha_1 * P_1 + \alpha_2 * P_2 + \alpha_3 * P_3 + \alpha_4 * P_4 + \alpha_5 * P_5$$

where P_1 is the maximum confidence value of all its conceptual relations; P_2 is the accuracy of the *concept* agents that are in relation with; P_3 expresses the accuracy of the conceptual relations of the *concept* agent; P_4 is the depth in the ontology; P_5 is the proportion of relevant *term* agents that denote this *concept* agent and $\alpha_1, \alpha_2, \alpha_3, \alpha_4, \alpha_5$ are the different weights of P_i. More precisely:

- P_1 = MaxConceptualRelationConfidence;
- P_2 = (nbRCAInTOR - nbRCANotInTOR)/(nbRCAInTOR + nbRCANotIn-TOR) where RCA (Related Concept Agent) are *concept* agents in relation with the evaluated *concept* agent;
- P_3 = (nbAcceptedCRA - nbRejectedCRA)/(nbAcceptedCRA + nbRejected-CRA) where CRA (Conceptual Relations of the Agent) are conceptual relations known by the evaluated *concept*;
- P_4 = {-1;1}: -1 if the *concept* agent is connected to the TOP agent of the ontology, 1 otherwise;
- P_5 = (nbRelevantTermAgent - nbNotRelevantTermAgent)/(nbRelevantTerm-Agent + nbNotRelevantTermAgent) where the *term* agents are denoting the *concept* agent.

After some experiments with DYNAMO-MAS we empirically fixed the values of α_1 to 0.5, α_2 to 1, α_3 to 1, α_4 to 1 and α_5 to 2.

Concept Agent Cooperative Behavior. The cooperative behavior of a *concept* agent is defined through three cooperative behaviors (according to the AMAS approach): two reorganization behaviors and an evolution behavior.

The first **reorganization behavior** is similar to the one of *term* agents. When a *concept* agent is connected to a valid *concept* agent, it informs its neighbors of this information. This helps *term* agents or *concept* agents to move in the AMAS.

Rule: When a *concept* agent is connected to another valid *concept* agent, it informs its neighbors of this information.

The second **reorganization behavior** concerns *concept* agent moving with transverse relations. To allow a *concept* agent to change its neighborhood thanks to its transverse relations, we took inspiration from the differentials principles [6]. These principles enable to identify one unit (or concept), its significance according to its identities (or similarities) and its differences with its neighbors.

In the DYNAMO project, each TOR is built around a core of stable concepts (2, 3 or 4 concepts) linked to the *DomainThing* concept agent by an *is_a* relation and connected amongst themselves by properties. So the addition of other concepts to the core concepts leads to the evolution of the TOR. For the AMAS, the goal of a *concept* agent is then to find under which core *concept* agent it has to be located. For that, when a *concept* agent has transverse relations from which it is the source or the target and when it is connected to the *DomainThing concept* agent, it moves in order to increase its relevance. For this, it chooses its transverse relation whose confidence is maximal. Then it compares this relation with the transverse relations of the core *concept* agents in order to choose the agent toward which to move. Finally, it sends a request to establish a *is_a* relation with a confidence equal to the confidence of the transverse relation with the concerned *concept* agent (❶) (Fig. 9). The latter accepts the establishment of the relation and sends a notification of acceptance (❷). Once the notification received, the *concept* agent changes its neighborhood by adding this new *concept* agent. This behavior can trigger a NCS if a *concept* agent has two or more transverse relations with the same confidence.

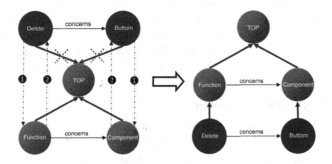

Fig. 9. Self-organization with transverse relations.

NCS8: A *concept* agent is facing an ambiguity NCS if it cannot choose between two or more transverse relations in order to change its neighborhood.

NCS8: To solve this NCS, a *concept* agent chooses the transverse relation whose end is a valid *concept* agent. If all the *concept* agents are valid or invalid, it chooses the relation with the *concept* agent that has the highest confidence. Finally, if all the confidences are equal, the *concept* agent chooses a arbitrary transverse relation. Thus, by changing its neighborhood, a *concept* agent increases its relevance and therefore increases its chance to propose itself to the ontologist.

The **evolution behavior** aims at preventing useless intermediate *concept* agents.

NCS9: An intermediate *concept* agent is facing a uselessness NCS if it has the same frequency[6] as its father *concept* agent and it has the same frequency as its son *concept* agent(s).

---

[6] The frequency of a *concept* agent is calculated from the sum of the frequencies of *term* agents that denote it.

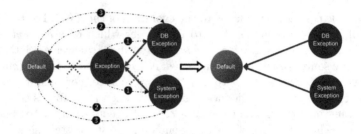

Fig. 10. Example of interaction for the simplification of the hierarchy.

NCS9 solution: A *concept* agent faced with a uselessness NCS disappears from the AMAS.

An intermediate *concept* agent informs its neighboring agents before disappearing in order to allow them to update their knowledge (❶) (Fig. 10). This disappearance can lead to a uselessness NCS. To avoid this, the *concept* agent proposes its father *concept* agent to its son *concept* agents as well as to its *term* agents that denote it (❶). Each agent then sends a request to establish a relation with this new *concept* agent (❷). The latter accepts the establishment of the relation and it sends a acceptance notification to the concerned agent (❸).

5.3   The Proposition Manager

The Proposition Manager enables the ontologist to visualize the ontology and the MAS proposals as well as to interact with the MAS (Fig. 3). Its main goals are (*i*) to sort out the proposals (to be displayed *via* the GUI) sent by the *concept* agents and the *term* agents; and (*ii*) to convey the corrections made by the ontologist to the *concept* and *term* agents with regard to their proposals.

Once the activity of the MAS is stabilized, i.e. when all the agents have processed the requests that they received, the Proposition Manager sorts the proposals, deletes contradictory ones and conveys the final list to the ontologist. Some contradictions may appear either when an agent proposes itself but later disappears from the MAS, or when an agent proposes a conceptual relation and later proposes another relation involving the same concepts with a different relation. The system considers that the most recent one is the only one that is valid.

The Proposition Manager finally sends back to agents that made proposals, the evaluation coming from the ontologist. This notification corresponds to a new request sent to the agents that will process it.

6   Evaluation of the DYNAMO-MAS

The DYNAMO-MAS was implemented as a plugin in the ontology editor Protégé[7] (Fig. 11). It completes TextViz [34], a tool dedicated for the semantic annotation of documents.

---

[7] http://protege.stanford.edu/

When the ontologist adds new texts to the corpus, the DYNAMO-MAS is triggered. The corpus analyzer extracts new candidate terms and new lexical relations. It sends them as inputs to the MAS; new term agents are created and interact with the other agents. So most of the agents in the MAS update their knowledge, react and communicate with each other until a stable state is reached. Then the MAS proposes a new ontology (the initial ontology that has been enriched and modified) to the ontologist in the GUI (Fig. 11) (❹). Changes are displayed in the concept panel (❶) and in the terms panel (❷). Proposed concepts and terms are the underlined one. A second Tab Widgets, called *DYNAMO - Virtual Ontologist Proposals* (❸), has been added to Protégé. It is a tabular view of the MAS proposals, incorporating non-hierarchical relations that cannot be seen in the two first panels. The ontologist validates, deletes or modifies the concepts, terms and relations proposed by DYNAMO-MAS *via* the Graphical User Interface (❹).

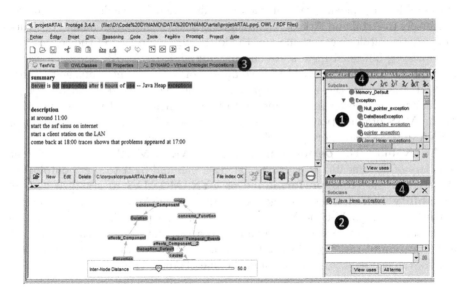

Fig. 11. The DYNAMO tool into the Protégé ontologies editor.

We tested our MAS on 3 different TOR and associated corpus provided by 3 partners of the DYNAMO project: (*i*) an English one on software bugs reports (made up of 887 terms and 582 concepts; the corpus is composed of 287 documents bug report files provided by Artal technology), (*ii*) a French one on automotive diagnosis (made up of 579 terms and 330 concepts; the corpus is composed of 710 documents files that report fault descriptions and repair procedures provided by Actia) and (*iii*) a French one on archaeology (made up of 733 terms and 380 concepts; the corpus is composed of 299 documents rule-based formulation of scientific papers provided by Arkeotek).

To evaluate the quality of DYNAMO-MAS we made a comparison between manual ontology evolution and automatic ontology evolution. The evolution is manual if the ontologist adds by itself new terms and new concepts in the ontology. The evolution is automatic, if the ontologist uses our tool. Obtained results can be seen in [37].

After the addition of 21 documents in the Artal corpus, DYNAMO-MAS provided **67 %** of term proposals and **56 %** of concept proposals that are relevant. It was also able to suggest 12 terms and 9 concepts not manually identified by the ontologist. After the addition of 12 documents in the Arkeotek corpus, DYNAMO-MAS provided **68.75 %** of relevant term proposals and **59.26 %** of relevant concept proposals. It was also able to suggest 18 terms and 14 concepts not identified. Finally, after the addition of 21 documents in the Actia corpus, DYNAMO-MAS reached only **16.98 %** of relevant term proposals and **22.22 %** of relevant concept proposals. It was also able to suggest 5 terms and 5 concepts not manually identified. These results are less good because new documents added to the corpora contained very little new knowledge. Furthermore, conversely to Artal and Arkeotek ontologies (that are ontologies under construction), Actia ontology covers most of the knowledge expressed in the corpus.

However, obtained results are encouraging and the AMAS approach seems relevant. Even if DYNAMO-MAS is composed of a large number of agents (more than 1000 agents), the new *concept* agents and *term* agents, with only local and distributed mechanisms, come up to find by themselves their best position in the MAS organization (the ontology). Results of the quality evaluation of DYNAMO-MAS are more precisely presented and discussed in [37].

To evaluate the performances of DYNAMO-MAS we added different numbers of documents (4, 8, 16, 32, 64, 128, 256 and 512) to the 3 ontologies. We studied the **time performance** of DYNAMO-MAS to stabilize and its **scalability**. Indeed, when new agents are added to the MAS, this latter becomes perturbed. The goal of the MAS is then to solve its perturbations in order to come back to a stable state and give rise to a new ontology. Initially, after the agentification of the 3 ontologies, the ontology of Artal was composed of 1469 agents, the one of Arkeotek 1328 and the one of Actia 1176.

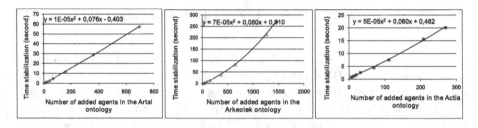

Fig. 12. Stabilization time of the MAS and number of added agents further to the addition of new documents.

Figure 12 shows the time required for the MAS to self-stabilize after the addition of new documents. The 3 curves having a X^2 coefficient close to zero, the time required for the MAS to stabilize is almost linear. The addition of 512 documents to the Arkeotek corpus has generated the creation of 1468 new agents in the MAS and the stabilization time took around 4 min and 30 s. The addition of 512 documents to the Artal corpus has generated the creation of 694 new agents and the stabilization time was around 1 min. Finally, the addition of 512 documents to the Actia corpus has generated the creation of 270 new agents and the stabilization time was around 20 s. We can conclude that DYNAMO-MAS stabilizes very quickly despite the large number of agents composing the MAS.

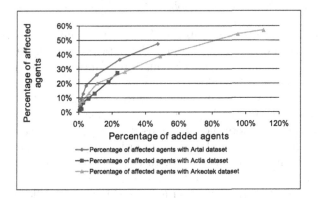

Fig. 13. Perturbation diffusion inside DYNAMO-MAS.

Figure 13 shows the percentage of perturbed agents after the introduction of new agents. In the 3 curves, we can notice that the addition of new agents to the MAS does not disrupt the whole system. When we added 512 documents to the Arkeotek corpus, the 1468 new agents (representing **110, 47 %** of the MAS) perturb only 755 agents (representing **56, 85 %** of the MAS). Also, when we added 512 documents to the Artal corpus, the 694 new agents (representing **47, 24 %** of the MAS) perturb only 696 agents (representing **47, 38 %** of the MAS). Finally, when we added 512 documents to the Actia corpus, the 270 new agents (representing **23, 14 %** of the MAS) perturb only 319 agents (representing **27, 34 %** of the MAS). This result shows that the disturbance can be considered as local.

As we seen, results provided by DYNAMO-MAS have a different quality depending on the corpus and on the initial state of the ontology. It seems that as the ontology construction progresses, the insertion of results supplied by DYNAMO-MAS becomes more complex, or conversely when the conceptualization is in its early stages, the MAS brings more help to the ontologist. In other words, the results of the qualitative evaluation of DYNAMO show that more an ontology is unachieved, more the AMAS proposals are accepted (Artal and Arkeotek data). Conversely, if an ontology is already built, the AMAS proposals are often rejected (Artal data). The completion of an ontology can be quantified thanks to the annotations of new documents added to the corpus. These values are calculated by TextViz and indicate the degree of annotation of a document by the ontology.

We are currently working on some improvements of DYNAMO-MAS in order to better adapt the results provided by the MAS to the ontologist, that is to say to personalize the system to the ontologist's actions. The idea here is to adjust the threshold proposals of agents depending on the state of completion of the ontology as well as according to the ontologist's actions.

We propose to take into account the results of annotation for the calculation of the proposals threshold.

$$InitialProposalThreshold = \frac{\sum_{i=1}^{n} annotationValue_i}{n}$$

This formula increases the proposals threshold when documents are rightly annotated and decreases otherwise. Thus, for the data provided by Actia, only the *term* agents and the *concept* agents having a relevance value higher than 9 will propose themselves to the ontologist.

Thereafter, the system has to learn how to vary this threshold according to the interactions with the ontologist. For that, we consider that more the ontologist accepts proposals more the un-proposed agents are potentially interesting. Thus, the threshold has to be decreased in order to propose them to the ontologist. Conversely, more the ontologist refuses proposals, more it makes sense to limit the number of proposals. The solution is then to increase the threshold. We propose to implement this adaptation mechanism with the following formula:

$$ProposalThreshold = ProposalThreshold - ProposalThreshold * Coef$$

Where ProposalThreshold is the current proposals threshold and agentRelevancy is the confidence of the accepted/refused/moved agent by the ontologist. The value of the *Coef* variable has to be different according to the ontologist's actions. It can have a value equal to 0.5 when the ontologist accepts a proposal (proposal at the right position in the TOR). It can have a value equal to 0.25 the ontologist moves a proposal (proposal correct but misplaced in the TOR).

When the ontologist refuses a proposal the ProposalThreshold can be defined with the following formula where *coef* can be equal to 0.5:

$$ProposalThreshold = ProposalThreshold + ProposalThreshold * Coef$$

Subsequently after a given number of the ontologist's actions, the Proposition Manager updates the proposals made by the ontologist by eliminating those that do not exceed the proposals threshold and by adding those that exceed it. A new updated version of the TOR is then proposed to the ontologist. Another perspective, is to introduce an Adaptive Value Tracker (AVT) component [27] to adjust the value of the *Coef* parameter. An AVT is a software component that finds the optimal value of a dynamic variable in a given space thanks to successive feedbacks. In our case, feedbacks come from the ontologist. When the ontologist rejects or accepts a set of concepts, relations and terms, the AVT will adjust *Coef*.

7   Conclusion

The aim of this paper was to propose a system to automatize the co-construction and the evolving (with an ontologist) of an ontology from texts. This issue is very interesting because manual construction and evolution of an ontology are complex and consuming tasks. The goal of our system was to automatically propose to the ontologist new concepts and/or terms to improve the ontology after the addition of new documents in the corpus. The ontologist can then make corrections on the ontology to be taken into account at runtime by the system in order to improve the ontology. After having proposed a state of art on this issue, and considering the characteristics of such issue, we justified why the Multi-Agent System paradigm and especially the Adaptive Multi-Agent System (AMAS) approach seem very relevant to resolve this problem. The AMAS approach proposes an original solution to solve complex problems evolving in a dynamic environment and for which an a priori know solution does not exist.

The core of this paper was to present the instantiation of this approach to the problem of co-construction and evolution of an ontology from texts with an ontologist. We gave in that sense a quick overview of the DYNAMO project in which this work took part as well as the architecture of the proposed system (DYNAMO-MAS). Then we focused on the cooperative behaviors we gave to the two types of agents we defined. In accordance with the AMAS approach, we put the emphasize on the Non Cooperative Situations to which each type of agents can be confronted with during the system functioning, as well as the actions they have to implement to come back to a cooperative state. Some experiments that were carried out within the project were then presented, with three ontologies more or less achieved, expressed in two different languages (French and English), concerning three different domains. These experiments confirmed that linguistic clues are not sufficient to decide the content of an ontology, and that the intervention of an ontologist is fundamental. We proposed then some improvements that are currently on-going in order to better co-construct the ontology, that is to say[8] to personalize[9] the results provided by DYNAMO-MAS with the actions of an ontologist (who can be more or less strict).

The originalities of our work are triple. First it enables to make evolve an ontology when new documents are added in the corpus without restarting the process of construction from scratch. Also, our proposition considers an ontology as a MAS that interacts with an ontologist and self-adapts when new domain knowledge are added. Finally our approach is independent of the language and the domain of the handled texts and it expresses the ontology in OWL format, a standard that makes it easily reusable.

Acknowledgments. We thank all the members of the DYNAMO project for their contribution and especially S. Rougemaille and M. Mbarki for their contribution to the implementation and the evaluation of the DYNAMO tool.

---

[8] Unsolved Problems for Emerging Technologies - Upetec company.

[9] Artal Technologies company.

References

1. Afsharchi, M., Far, B.H.: Automated ontology evolution in a multi-agent system. In: 1st International Conference on Scalable Information Systems, InfoScale '06, New York, NY, USA. ACM (2006)
2. Akinsola, T.M.: Automated ontology evolution. Masters of Science Informatics, University of Edinburgh, Edinburgh, Scotland (2008)
3. Aldea, A., Bañares-alcántara, R., Bocio, J., Gramajo, J., Isern, D.: An ontology-based knowledge management platform. In: Workshop on Information Integration on the Web Associated to IJCAI, pp. 177–182 (2003)
4. Aubin, S., Hamon, T.: Improving term extraction with terminological resources. In: Salakoski, T., Ginter, F., Pyysalo, S., Pahikkala, T. (eds.) FinTAL 2006. LNCS (LNAI), vol. 4139, pp. 380–387. Springer, Heidelberg (2006)
5. Aussenac-Gilles, N., Hernandez, N.: Du linguistique au conceptuel: identification de relations conceptuelles à partir de textes. In: Atelier "Acquisition et modélisation de relations sémantiques, Toulouse (2009)
6. Bachimont, B.: Engagement sémantique et engagement ontologique: conception et réalisation d'ontologies en ingénierie des connaissances. Ingénierie des Connaissances: Evolutions récentes et nouveaux défis 1, 1–16 (2000)
7. Bao, J., Honavar, V.: Collaborative ontology building with wiki@nt. In: Workshop on Evaluation of Ontology-Based Tools (2004)
8. Bergenti, F., Poggi, A., Rimassa, G., Turci, P.: Comma: a multi-agent system for corporate memory management. In: International Joint Conference on AAMAS, pp. 1039–1040 (2002)
9. Bernon, C., Capera, D., Mano, J.-P.: Engineering self-modeling systems: application to biology. In: Artikis, A., Picard, G., Vercouter, L. (eds.) ESAW 2008. LNCS, vol. 5485, pp. 248–263. Springer, Heidelberg (2009)
10. Buitelaar, P., Cimiano, P., Magnini, B.: Ontology Learning from Text: Methods, Evaluation and Applications. Frontiers in Artificial Intelligence and Applications Series. IOS Press, Amsterdam (2005)
11. Camps, V.: Vers une théorie de l'auto-organisation dans les systèmes multi-agents basée sur la coopération: application à la recherche d'information dans un système d'information répartie. Ph.D. thesis, Université Paul Sabatier, Toulouse, Janvier 1998
12. Chagnoux, M., Hernandez, N., Aussenac-Gilles, N.: An interactive pattern based approach for extracting non-taxonomic relations from texts. In: Workshop on Ontology Learning and Population (Associated to ECAI 2008), pp. 1–6. University of Patras, Juillet 2008
13. Cimiano, P.: Ontology Learning and Population from Text: Algorithms, Evaluation and Applications. Springer, Boston (2006)
14. Cimiano, P., Völker, J.: Text2Onto - a framework for ontology learning and data-driven change discovery. In: Montoyo, A., Muñoz, R., Métais, E. (eds.) NLDB 2005. LNCS, vol. 3513, pp. 227–238. Springer, Heidelberg (2005)
15. Elmore, M.T., Potok, T.E., Sheldon, F.T.: Dynamic data fusion using an ontology-based software agent system. In: 7th World Multiconference on Systemics, Cybernetics and Informatics (2003)
16. Flouris, G.: On belief change and ontology evolution. Ph.D. thesis, Department of Computer Science, University of Crete, Heraklion, Greece (2006)
17. Flouris, G., Plexousakis, D., Antoniou, G.: A classification of ontology change. In: CEUR-WS 201. CEUR-WS.org (2006)

18. Gandon, F.: Distributed artificial intelligence and knowledge management: ontologies and multi-agent systems for a corporate semantic web. Thèse de doctorat, Université de Nice - Sophia Antipolis, Novembre 2002
19. Gawrysiak, P., Protaziuk, G., Rybiński, H., Delteil, A.: Text onto miner – a semi automated ontology building system. In: An, A., Matwin, S., Raś, Z.W., Ślezak, D. (eds.) Foundations of Intelligent Systems. LNCS (LNAI), vol. 4994, pp. 563–573. Springer, Heidelberg (2008)
20. Gleizes, M.-P., Camps, V., Georgé, J.-P., Capera, D.: Engineering systems which generate emergent functionalities. In: Weyns, D., Brueckner, S.A., Demazeau, Y. (eds.) EEMMAS 2007. LNCS (LNAI), vol. 5049, pp. 58–75. Springer, Heidelberg (2008)
21. Greenwood, D., Lyell, M., Mallya, A., Suguri, H.: The IEEE FIPA approach to integrating software agents and web services. In: AAMAS (2007)
22. Hadzic, M., Dillon, D.: An agent-based data mining system for ontology evolution. In: Meersman, R., Herrero, P., Dillon, T. (eds.) OTM 2009 Workshops. LNCS, vol. 5872, pp. 836–847. Springer, Heidelberg (2009)
23. Harris, Z.S.: Mathematical Structures of Language. Wiley, New York (1968)
24. Hearst, M.A.: Automatic acquisition of hyponyms from large text corpora. In: 14th International Conference on Computational Linguistics, pp. 539–545 (1992)
25. Klein, M.: Change management for distributed ontologies. Ph.D. thesis, Dutch Graduate School for Information and Knowledge Systems, Germany (2004)
26. Leen-Kiat, S.: Multiagent distributed ontology learning. In: Workshop on Ontologies in Agent Systems, associated to AAMAS, Bologna, Italy, July 2002, vol. 66, pp. 75–79 (2002)
27. Lemouzy, S.: Systèmes interactifs auto-adaptatifs par systèmes multi-agents auto-organisateurs: application à la personnalisation de l'accès à l'information. Thèse de doctorat, Université Paul Sabatier, Toulouse, Juillet 2011
28. Levenshtein, V.I.: Binary codes capable of correcting deletions, insertions and reversals. Soviet Physics Doklady 10, 707 (1966)
29. Maedche, A.: Ontology Learning for the Semantic Web, vol. 665. Springer/Kluwer Academic Publisher, Boston (2002)
30. Ottens, K.: Un système multi-agent adaptatif pour la construction d'ontologies à partir de textes. Thèse de doctorat, Université Paul Sabatier, Toulouse, Octobre 2007
31. Ottens, K., Hernandez, N., Gleizes, M.-P., Aussenac-Gilles, N.: A multi-agent system for dynamic ontologies. J. Log. Comput. (Special Issue on Ontology Dynamics) 19, 1–28 (2008)
32. Picard, G., Gleizes, M.-P.: Cooperative self-organization: designing robust and adaptive robotic collectives. In: Third European Workshop on Multi-Agent Systems, Brussels, Belgium, pp. 495–496. KVAB, Brussel (2005)
33. Reinberger, M.-L., Spyns, P.: Discovering knowledge in texts for the learning of dogma-inspired ontologies. In: Workshop on Ontology Learning and Population, ECAI04, Valencia, pp. 19–24 (2004)
34. Reymonet, A., Thomas, J., Aussenac-Gilles, N.: Modelling ontological and terminological resources in OWL DL. In: OntoLex07 - associated to ISWC, Busan (2007)
35. Safari, L., Afsharchi, M., Far, B.H.: Concepts in action: performance study of agents learning ontology concepts from peer agents. In: ICAART'09, pp. 526–532 (2009)
36. Sellami, Z.: Gestion dynamique d'ontologies à partir de textes par systèmes multi-agents adaptatifs. Thèse de doctorat, Université de Toulouse, Juillet 2012

37. Sellami, Z., Camps, V., Aussenac-Gilles, N.: DYNAMO-MAS: a multi-agent system for ontology evolution from text. J. Data Semant. **2**(2), 145–161 (2013). doi:10.1007/s13740-013-0025-1

38. Sellami, Z., Camps, V., Aussenac-Gilles, N., Rougemaille, S.: Ontology Co-construction with an adaptive multi-agent system: principles and case-study. In: Fred, A., Dietz, J.L.G., Liu, K., Filipe, J. (eds.) IC3K 2009. CCIS, vol. 128, pp. 237–248. Springer, Heidelberg (2011)

39. Séguéla, P.: Construction de modèles de connaissances par analyse linguistique de relations lexicales dans les documents techniques. Thèse de doctorat, Université Paul Sabatier, Toulouse, Mars 2000

40. Siebes, R., van Harmelen, F.: Ranking agent statements for building evolving ontologies. In: Workshop on Meaning Negotation, in Conjunction with the Eighteenth National Conference on Artificial Intelligence, July 2002

41. Slimani, S., Baina, S., Baina, K.: A framework for ontology evolution management in SSOA-based systems. In: IEEE International Conference on Web Services, pp. 724–725 (2011)

42. Slimani, S., Baïna, S., Baïna, K.: Interactive ontology evolution management using mutli-agent system: a proposal for sustainability of semantic interoperability in SOA. In: WETICE, pp. 41–46 (2011)

43. Stojanovic, L.: Methods and tools for ontology evolution. Ph.D. thesis, Karlsruhe University, Germany (2004)

44. Tamma, V., Bench-Capon, T.: An ontology model to facilitate knowledge-sharing in multi-agent systems. Knowl. Eng. Rev. **17**, 41–60 (2002)

45. Velardi, P., Navigli, R., Cucchiarelli, A., Neri, F.: Evaluation of ontolearn, a methodology for automatic learning of domain ontologies. In: Buitelaar, P., Cimiano, P., Magnini, B. (eds.) Ontology Learning from Text: Methods, Evaluation and Applications. IOS Press, Amsterdam (2005)

46. Viollet, A.: Un protocole entre agents pour l'alignement d'ontologies. Université Joseph Fourier, Grenoble, Rapport de master (2004)

47. Wang, J., Les Gasser, L.: Mutual online ontology alignment. In: Workshop on Ontologies in Agent Systems, associated to AAMAS, Bologna, Italy, July 2002, vol. 66, pp. 103–113 (2002)

48. Xuan, D.N., Bellatreche, L., Pierra, G.: Un modèle à base ontologique pour la gestion de l'évolution asynchrone des entrepôts de données. Modélisation. Optimisation et Simulation des Systèmes: Défis et Opportunitès), Rabat, Maroc, pp. 1682–1691 (2006)

49. Zablith, F., Sabou, M., d'Aquin, M., Motta, E.: Ontology evolution with evolva. In: Aroyo, L., Traverso, P., Ciravegna, F., Cimiano, P., Heath, T., Hyvönen, E., Mizoguchi, R., Oren, E., Sabou, M., Simperl, E. (eds.) ESWC 2009. LNCS, vol. 5554, pp. 908–912. Springer, Heidelberg (2009)

50. Zablith, F., Sellami, Z., D'Aquin, M., Aussenac-Gilles, N., Hernandez, N.: Vers la combinaison de deux techniques d'évolution d'ontologies à partir de ressources générales et de ressources linguistiques. In: Atelier Evolution d'ontologies des 21e Journées francophones d'Ingénierie des Connaissances, Nîmes 2010

Game Theoretical Model for Adaptive Intrusion Detection System

Jan Stiborek[1,2](✉), Martin Grill[1,2], Martin Rehak[1,2], Karel Bartos[1,2], and Jan Jusko[1,2]

[1] Agent Technology Center, Department of Computer Science, Czech Technical University in Prague, Prague, Czech Republic
[2] CISCO Systems, Inc., San Jose, USA
{jastibor,magrill,marrehak,kbartos,jajusko}@cisco.com

Abstract. We present a self-adaptation mechanism for network intrusion detection system based on the use of game-theoretical formalism. The key innovation of our method is a secure runtime definition and solution of the game and real-time use of game solutions for immediate system reconfiguration. Our approach is suited for realistic environments where we typically lack any ground truth information regarding traffic legitimacy/maliciousness and where the significant portion of system inputs may be shaped by the attacker in order to render the system ineffective. Therefore, we rely on the concept of challenge insertion: we inject a small sample of simulated attacks into the unknown traffic and use the system response to these attacks to define the game structure and utility functions. This approach is also advantageous from the security perspective, as the manipulation of the adaptive process by the attacker is far more difficult.

1 Introduction

Detection of intruders, either humans or automated solutions such as malware or botnets in computer networks is one of the key problems of modern computer security field. The attackers are getting increasingly sophisticated, using efficient illegal marketplaces to build ad-hoc collaborations between specialists in vulnerability identification, malware (malicious software) writing, malware propagation and botnet building and finally in using the malware to attack individuals or companies.

This paper presents a game theoretical model of adaptation processes inside an autonomic, self-optimizing Intrusion Detection System (See *Appendix* for details about the IDS solution used). Our goal is first and foremost to analyze the risks related to opponent's manipulation of system internal state and configuration, performed in order to reduce its effectiveness. This addresses the existing concern with expected increase in malware sophistication - theoretical models for distributed learning in malware exist [1], and strategic manipulation of Intrusion Detection Systems by shaping of the input data has been demonstrated,

© Springer-Verlag Berlin Heidelberg 2014
R. Kowalczyk et al. (Eds.): TCCI XV, LNCS 8670, pp. 133–163, 2014.
DOI: 10.1007/978-3-662-44750-5_7

albeit offline [2]. This behavior corresponds to wider context of targeted attacks on learning processes, studied in the fields of adversarial machine learning and adversarial classification [3].

Therefore, if we want to introduce a new layer of environment-driven adaptation into the intrusion detection system [4], we need to ensure what is the extent to which can the opponent misuse the reconfiguration layer to reduce system's effectiveness.

The principal question this paper investigates is simple: What is the cost of preventive IDS resistance to the attackers with access to internal state information and outputs of an IDS, in terms of suboptimal False positives/False Negative values? In other words, we measure whether and by how much will the IDS reconfiguration against the "worst case", highly sophisticated attacks with insider access reduce its performance against the "standard", relatively unsophisticated attackers with no knowledge of IDS existence, nominal effectiveness and current internal state.

In order to answer the above question, we use the methods from the field of game theory [5] and decision theory. These concepts, introduced in Sect. 3 are mapped to IDS structure in Sect. 3.1. They conceptualize the relationship between the attacker and the defender as a two party, non-zero sum game, where the attack/defence actions of both players correspond to strategies in the game-theoretical model of their interaction.

Note that presented solution was successfully deployed in real-world scenario in IDS system CAMNEP where it serves as one of key components. Results of the research presented in this paper were successfully transferred into industry by technological company Cognitive-Security s.r.o. The importance of this work was recently confirmed by acquisition of this company by CISCO Systems, Inc. – one of the technological leaders in network security.

2 Related Work

Our work belongs into the broader field of regret minimization techniques [6]. This is due to the fact that the algorithm is deployed online and makes decision based on partial and biased information. These decision are then evaluated *ex post*, and we can determine the difference between the actually achieved utility and maximally achieved utility, given a fixed set of strategies to select from (external regret). The use of regret minimization techniques based on explicit strategic reasoning is novel in the intrusion detection field. Traditionally, most of the work uses the game theoretical models only for formal analysis of highly simplified scenarios [7–10]. Main issue is that such simple models do not incorporate the dynamic structure of protected network and thus do not optimize the strategy against current state of the network. In [11] authors propose game theoretical framework that is able to learn policies or configurations of an IDS system in iterative manner. Such approach is clearly feasible and provides optimal results but proposed model consider only the current configuration as the state of the system and does not capture the influence of the background network traffic that heavily affects the performance of an IDS system especially in the case of anomaly-based IDS.

Another important inspiration point is the use of agent-based modeling and game theoretical techniques for physical security problems, such as in guard dispatch at LAX [12], where the techniques used are analogous to ours, except for non-stackelberg nature of our problem. On the side of the attackers [1], the design of learning and evolving malware makes the problem of strategic evasion important, as the learning algorithms can effectively participate as strategically behaving players, and can use fairly advanced techniques mentioned above [2,3].

The seminal work of Alpcan [7] analyzes the IDS game as a sequence of interactions between strategically reasoning opponents and a network of IDS sensors.

However, the model used in the Alpcan's paper fails to capture some of the important aspects, such as problem dynamic nature, more realistic utility functions and a necessary overlap between the detection domains of multiple sensors on the network. In [8], Alpcan and Basar extend the above model considerably. Their formalism, based on a combination of Markov games and Q-learning, actually links the agent's performance as a detector/learner to its game performance by representing the imperfect information.

3   IDS Game Model

In this paper, we will use the simplest model available in the field of the game theory, a single stage game of two players. Each such game is defined as a three tuple:

$$G = (P, S, U) \tag{1}$$

- where P is a set of **players** traditionally indexed as $P = 1, 2$, in our case denoted $P = d, a$, where the player a is the attacker (column player) and the player d is the defender (row player),
- S is a set of **strategies** available to all players. In our case, where the strategies are disjunctive, we impose simply $S = d_1, ..., d_i, ..., d_m, a_1, ..., a_j, ..., a_n$, here the strategies d_i are those of the defender and the strategies a_i are available to the attacker, and
- U denotes **utility function** of the form: $U : S \times S \rightarrow \mathbb{R} \times \mathbb{R}$, or less formally: $U : d_i \times a_i \rightarrow (u_d, u_a)$. Utility function returns the game payoff of the defender u_d and the attacker u_a when these invoke the strategies d_i and a_i respectively. Payoffs are real valued, and are frequently negative. Note that the negative value of payoff signifies the loss for the player, and unlike in the case of zero-sum games, this loss does not become other player's gain. The game structure can be alternatively defined by two matrices that link the

attacker's and defender's strategies with the payoff functions for each player[1]. Such alternative definition is defined in Eq. 2 for defender and 3 for attacker.

$$
u_d = \begin{pmatrix}
Def./Att. & a_1 & a_2 & a_3 & \cdots & a_n \\
d_1 & u_d(d_1, a_1) & u_d(d_1, a_2) & u_d(d_1, a_3) & \cdots & u_d(d_1, a_n) \\
d_2 & u_d(d_2, a_1) & u_d(d_2, a_2) & u_d(d_2, a_3) & \cdots & u_d(d_2, a_n) \\
\vdots & \vdots & \vdots & \cdots & \ddots & \cdots \\
d_m & u_d(d_m, a_1) & u_d(d_m, a_2) & u_d(d_m, a_3) & \cdots & u_d(d_m, a_n)
\end{pmatrix}
\tag{2}
$$

$$
u_a = \begin{pmatrix}
Def./Att. & a_1 & a_2 & a_3 & \cdots & a_n \\
d_1 & u_a(d_1, a_1) & u_a(d_1, a_2) & u_a(d_1, a_3) & \cdots & u_a(d_1, a_n) \\
d_2 & u_a(d_2, a_1) & u_a(d_2, a_2) & u_a(d_2, a_3) & \cdots & u_a(d_2, a_n) \\
\cdots & \cdots & \cdots & \cdots & \cdots & \cdots \\
d_m & u_a(d_m, a_1) & u_a(d_m, a_2) & u_a(d_m, a_3) & \cdots & u_a(d_m, a_n)
\end{pmatrix}
\tag{3}
$$

The gameplay of this game type is very simple in our case: both players simultaneously select their strategies from the set S and the combination of these strategies determines the payoffs to attacker and defender, as defined by their respective utility functions. Note that due to the inherent nature of the IDS problem, the game is not a zero sum one, where the gain of one player results in the equivalent loss of the other player – combination of strategies therefore affects not only the distribution of payoffs between the players, but also the total sum of payoffs.

3.1   Intrusion Detection Game Specification

This game structure introduced above allows us to reason about the outcome of the interaction between the attacker and the defender, and potentially identify the likely outcomes of the game. Below, we will present the details of player's strategies, utility functions and solution concepts that will influence the outcome of the game.

Strategies. The pure strategy sets of both players form the following set

$$S = d_1, ..., d_i, ..., d_m, a_1, ..., a_j, ..., a_n$$

as defined in Eq. 1. The strategies $d_1, ..., d_i, ..., d_m$ are the strategies of the defender (row player), while the strategies $a_1, ..., a_j, ..., a_n$ are available to the attacker.

The defender's **pure strategies** are defined as a selection of one system configuration from a finite number of available configurations - playing the game is therefore functionally equivalent with the trust-based optimization described

---

[1] These two matrices can be collapsed into a single one in case of the zero sum game, where the gain of one of the player is directly translated into the equivalent loss of the other player.

in [4], where we have introduced an online regret-minimization mechanism suitable for dynamic and unstable environments.

When the defender plays a **mixed strategy**, it constructs a probability distribution over the set of pure strategies $d_1, ..., d_i, ..., d_m$, assigning a probability in the $[0, 1]$ interval to each pure strategy. The sum of probabilities (weights) of individual strategies must be 1. In practice, the mixed strategies will typically have a restricted support, with roughly 2–5 pure strategies with non-zero probability.

The attacker's pure strategies are defined even more easily. Each attacker's strategy is defined by performing one attack such as horizontal scan, vertical scan, host fingerprinting, buffer overflow, denial of service and others. Mixed strategies are defined similarly to defender, as a probability distribution on the support of attack actions. In practice, the attackers also execute their strategies in a particular, logical orderings (plans), but the identification and use of this behavior is outside of the scope of this paper.

Utility Functions. In contrast to previous work in IDS modeling [7–10], the utility functions that we use to represent player's gains and losses are not simplified, but are designed to provide a realistic model of incentives in real IDS system. This is made possible by the fact that we don't attempt to perform a formal analysis of the problem (even if we are still able to identify and verify several key properties of the system), but rather concentrate on online discovery of game parameters and runtime solution of the game in the context of specific threat environment and network traffic situation.

The form of the utility functions determines the characteristics of the game – if the game is a **zero sum game**, i.e. the sum of utilities of all players is constant over any combination of played strategies, it is relatively easy to identify stable Nash equilibria and other solutions, as we will discuss in Sect. 3.2. However, in our case, the game is not zero sum. This is a natural corollary of the criminal character of the activities [13] – crime can be commonly defined as an activity that redistributes the utility between the players at the expense of the overall utility reduction.

The utility functions in our game are relatively complex, as they need to reflect the complexity of the problem. The parameters of both utility functions are:

- $\alpha_{i,j}$ denotes the probability that the attack strategy a_j is detected when the defender selects the defence strategy d_i. Intuitively, in our case, it estimates the probability of the given detection strategy (i.e. aggregation function), combined with current status of detection agent's and the background traffic will be able to successfully raise an alarm upon the occurrence of an attack from the class corresponding to a_j.
- β denotes the probability that a given detection strategy, combined with current system state and background traffic, will result in a false positive. Note that this element is only present in defender's utility matrix, and its

manipulation can be used by the attacker to launch denial-of-service attacks on detection mechanisms [14].

- γ_j denotes the probability of attack success. It discounts the value of undetected breach from both the attackers and defenders standpoint, and typically features relatively low values.
- $V(t)$ denotes the background traffic volume that is used to estimate the number of false positives in combination with the parameter β.
- $P_d(a_j)$ denotes the defender's payoff/loss on attack success. It is most often a negative value, except for multi-tier honeypot systems or very particular situations where the defender can gain knowledge from the attacker bypassing the IDS. The loss can be relatively low in case of exploratory activities (scan, fingerprinting), but is relatively high when the attacker actually breaches a system. The attack tree-based methodology introduced in [4] allows the distribution of this ultimate utility between the pre-condition actions, effectively motivating the system to concentrate on all relevant stages of the attack (equivalent to attack classes a_j) during the game.
- $P_a(a_j)$ denotes the expected utility the attacker receives upon successful realization of given attack action from the attack class corresponding to strategy a_j.
- $D_a(a_j)$ denotes the attacker's payoff/loss on detection, which is typically a negative value. The value of this parameter can vary widely, as it can be very high for last stages of elaborate attacks executed inside defender's perimeter, or can be almost zero in case of internet attacks.
- $D_d(a_j)$ denotes the defender's payoff for attack detection. This parameter value is the main cause for the game not being a zero sum game in a general case, as the payoff is typically zero in enterprize or internet settings following the similar reasoning as in the $P_d(a_j)$ case – damages from the attacking party are almost impossible to seek (not even considering the problems related to the root attacker identification and burden of proof).
- $C_a(a_j)$ denotes the cost of the attack performance on the part of attacker. Typically very low for enterprize/internet-originating attacks.
- C_{TP} - denotes the (average) cost of processing of each detected incident (true positive) for the defender.
- C_{FP} - denotes the average cost of a false alarm for the defender, used in conjunction with β and $V(t)$ to estimate the false positive cost.
- C_M - denotes the fixed cost of monitoring infrastructure, independent on attack or traffic intensity.

The utility function of the defender has three principal components: the first term deals with successfully detected attacks, the second term represents the loss associated with undetected attacks and the third term describes the overhead of the monitoring, which consists of the false positive costs and the fixed cost of monitoring.

Individual utility functions are defined as follows. Defender's utility is:

$$u_d(d_j, a_i, t) = \alpha_{i,j}(D_d(a_j) - C_{TP}) + (1 - \alpha_{i,j})\gamma_j P_d(a_j)$$
$$- \beta V(t)C_{FP} - C_M \tag{4}$$

Attacker's utility can be described as:

$$u_a(d_j, a_i) = \alpha_{i,j}D_a(a_j) + (1 - \alpha_{i,j})\gamma_j P_a(a_j) - C_a(a_j) \tag{5}$$

Utility function terms. The first situation that we represent corresponds to attack detection. The term in the defender's utility function describing this situation (without the fixed costs of monitoring and false positives, which will be discussed below) is:

$$\alpha_{i,j}(D_d(a_j) - C_{TP}) \tag{6}$$

We can see that the defender may get some payoff from attack detection, but globally, the value of the term $D_a(a_j)$ would be zero or negative due to the reinstallation and recovery costs. The term C_{TP} represents the immediate cost of incident detection, investigation and processing.

On the attacker's side, this situation is described by the term:

$$\alpha_{i,j}D_a(a_j) \tag{7}$$

We can see that the loss of the attacker depends almost entirely on the impact the detection has on attacker's plans – the value can be relatively high in the last stages of complex attack plans deep within the protected perimeter, but is next to zero for malware propagation on Internet due to the lack of effective enforcement.

The second term of the utility function covers the situation when the attacks are not detected.

In the defender's case, the term is:

$$(1 - \alpha_{i,j})\gamma_j P_d(a_j) \tag{8}$$

The first factor corresponds to non-detection probability, while the factor γ_j describes the likelihood of attack/exploit success, which then amortizes the value of the successful execution $P_d(a_j)$. The impact of the factor γ_j is crucial. When there is an attack, the actual *defender's optimum in most situations is that the attack is both undetected and unsuccessful.* Reasoning behind this analysis is straightforward: Eq. 6 typically has the term $D_d(a_j)$ zero or negative[2],

---

[2] The only situations where this term is actually globally positive are those where an efficient counter-attack mechanism (in tactical/military problems) or intelligence-processing mechanism allows the defender to counter-attack the attacker's resources or to deduce attacker's goals, plans or at least intentions. From the other side of the problem, the attacker needs to structure its actions in such a way, that their eventual compromise would not give away disproportionally high volume of information about its goals or resources. This consideration is integrated in the value of the term $D_a(a_j)$.

the term $-C_{TP}$ is also negative and the best strategy is therefore to avoid detection of attacks with low loss/likelyhood product $\gamma_j P_d(a_j)$.

From the attacker's perspective, the second term is a straightforward amortization of success payoff by success likelihood and non-detection probability. Failure to exploit (with probability $1 - \gamma_j$) is also preferable to detection for the attacker, but only by the slight margin of the term $D_a(a_j)$, which is typically low for external attacks:

$$(1 - \alpha_{i,j})\gamma_j P_a(a_j) \tag{9}$$

The conclusion that some (i.e. unsuccessful) attacks are better left undetected may seem surprising, but it actually corresponds to very natural equilibria, due to the costs associated with any detected attacks. The problem in this case is therefore how to optimize the sensitivity of the intrusion detection system, so' that it will only detect the relevant threats. We attempt not only to remove the false positives, but also discount the value of true positives with little relevance to the actual system. This behavior ensures more effective monitoring with little or no impact on security.

The last component of both utility functions captures the costs related to cyber-attack or defense. Attacker's side lost utility can be trivially described as the cost associated with attack performance:

$$- C_a(a_j) \tag{10}$$

The defender's utilities depend on two principal factors: cost of the monitoring infrastructure and the cost of the processing of false positives, which can be significant for real world systems:

$$- \beta V(t)C_{FP} - C_M \tag{11}$$

The first of the two components estimates the number of false positives $(\beta V(t))$ and the total cost of their assessment $(\beta V(t)C_{FP})$, while the second term captures the fixed cost of monitoring, such as the infrastructure cost and fixed operation costs.

The size of these two terms is non-negligible – the number of false positives can rival the number of real incidents in open networks (see the Experimental section for more details), and false positives would typically significantly outnumber the true positives on internal, well-managed networks. These two terms are also the main reason why the IDS game is not a zero sum game, as they introduce a fundamental non-efficiency into the system.

On the other hand, when we can limit the number of false positives (hypothetically) and when we consider the value of the parameters C_M and $C_a(a_j)$ as negligible, we can hypothetically obtain a zero-sum game if the other parameters are tuned to ensure that the attacker's gains are matched by defender's losses and vice versa. An example of such scenario can be a tactical cyber-warfare, where the discovery of an electronic attack strategy can give away the attacker's intent, or can allow the defender to counter-attack the attacker's own network by exploiting the attacker's attack code faults/vulnerabilities or by providing disinformation. The importance of zero-sum games in these situations lies in their (comparatively) easy identification of Nash equilibria, and therefore in possibly more efficient response.

It is important to note that in typical cyber-attack scenarios, the game is actually highly asymmetric, as the attacker's costs $C_a(a_j)$ and potential losses are almost zero in the individual attack case, and the defender's much higher losses are amplified by relatively high cost of monitoring and false positives processing as specified in Eq. 11.

3.2 Solution Concepts

The definition of the game alone does not allow the player to identify the optimal behavior. There are several well-accepted solution concepts, based on different criteria of optimality. The ones that we have considered are the most commonly used ones:

– **Strategy dominance.** Dominant strategy is the simplest and strongest solution concept, where one of the strategies dominates another strategy or all other strategies. Formally, we say that the strategy a **dominates** the strategy b *iff* playing a guarantees at least the same payoff than playing b for any possible strategy of the opponent. In our case, we will write it in the form valid for the defender:

$$d* \text{ dominates } d' \text{ iff } \forall j : u_d(d*, a_j) \geq u_d(d', a_j)$$
$$\text{and } \exists j : u_d(d*, a_j) > u_d(d', a_j) \tag{12}$$

Strict domination, a stronger concept is defined as follows: Strategy a **strictly dominates** the strategy b *iff* playing a guarantees strictly better payoff than playing b for any possible strategy of the opponent.
Again, for sake of clarity, we will write it in the form valid for the defender:

$$d* \text{ dominates } d' \text{ iff } \forall j : u_d(d*, a_j) > u_d(d', a_j) \tag{13}$$

Strictly dominant strategy is then defined as a strategy that strictly dominates all other strategies. Weakly **dominant strategy** is a strategy that dominates all other strategies.
The downside of this solution concept is that under normal circumstances, we can only rarely find a strategy that dominates all the others, as we will see in the experimental results presented in Sect. 5.

– **Max-min rule.** This solution concept (similar to Minmax rule) selects the strategy with the highest minimal payoff for the player. This solution concept is a security strategy, and is especially relevant in the situations where we suspect that the opponent has access to the part of the system's internal state or even to strategy selection decision. Playing max-min (for the defender) covers the risk of the opponent playing the most damaging action:

$$d* = \arg\max_{d_i} \min_{a_j} u_d(d_i, a_j) \tag{14}$$

While being a relatively strong solution concept, it shall be noted that the max-min criteria, as well as the dominance criteria does not require any knowledge of the opponent's utility function – the strategy selected depends only on the defender's utility functions alone. This will not be true for the two last concepts introduced below.

- **Conditional dominance.** Conditional dominance [15] is a solution concept which is stronger than looking for dominant strategy, as it allows the players to use the information about the utility functions of both players to identify the set of strategies that are not dominated. Briefly, the concept is similar to dominance, but with the strategies of other players restricted to a specific subset. Therefore we use the term conditionally dominated to emphasize that the dominance is valid only on a subset of opponent's strategy spaces. Again, from the defender's perspective, the strategy d_i *is conditionally dominated* given the attacker's strategy set $a_1, ..., a_p$, where $p \leq m$ iff:[3]

$$\exists d_l, \ l \neq i, \ \forall a_k, \ k \leq p \ : u_d(d_i, a_k) < u_d(d_l, a_k) \qquad (15)$$

In our case, we use the conditional dominance (or rather non-dominance) iteratively for both players, in order to reduce the set of strategies rationalizable for players. Iterative application of the rule gives us a relatively large set of strategies (including mixed strategies) for both players, and both players randomly select the strategy from the set. This solution concept variant covers the very general case where we have a reasonably good information about opponent's means and strategies, but only a very limited information about its goals, as expressed by the form of its utility function. In short, it is only a baseline benchmark concept that we don't expect to be useful in any real settings.

- **Nash equilibrium.** The Nash equilibrium is a strong solution concept that identifies a stable combinations of player's strategies. It is defined as a state where no player can improve its payoff by unilaterally changing its strategy. Formally, the equilibrium, defined as a pair of strategies (either pure or mixed) of both players needs to fulfill the following condition:

$$(d_i, a_j) \text{ is a NE iff } \forall d_l, l \neq i : u_d(d_l, a_j) \leq u_d(d_i, a_j)$$
$$\text{and } \forall a_k, k \neq j : u_a(d_i, a_k) \leq u_a(d_i, a_j) \qquad (16)$$

where d_i and a_j are to be considered as mixed strategies. The major difference with the max-min rule is the number of equilibria solving the condition 16. It can be shown that when we admit solutions in mixed strategies, the IDS game as specified in this section always has at least one Nash equilibrium. However, we are typically able to identify more than one equilibrium in the game, and the players are then confronted with the problem which one to select (by playing the corresponding d_i or a_j strategy).

---

[3] If $p = n$, the concept is trivially equivalent to strict dominance. For sake of notation clarity, we arbitrarily select the first p attacker strategies, with no loss of generality.

The performance of solution concepts will be analyzed in Sect. 5, where we will compare them on both the challenge data[4], and also on their ability to handle a real world instance of an actual attack scenario. To understand the results, we need to understand the difference between the various types of optimality criteria in the system [6]:

- **Optimal strategy** is the player's best **pure** strategy from the set S in the time t. This is an **a-posteriori** concept, which can be only determined after once the execution was completed and the system results against the **actual** network attacks were determined. This value is independent of the solution concept used.
- **Optimal decision** is the strategy (mixed or pure) identified by the player as optimal a-priori, given the inputs available **a-priori** in the moment that the decision is taken. Making the optimal decision does not guarantee actually selecting the optimal strategy, principally for two reasons: information about the state of the system are incomplete/limited/biased (making the optimal strategy not present in the optimal decision, or decreasing its selection probability in the mix), or the pure strategy selected stochastically from the optimal decision was not actually the optimal strategy. Optimal decision can be obtained by the use of any of the four solution concepts described above (or any other solution concepts, such as trust-based mechanism), as each of the solution concepts introduces a slightly different bias into the optimality criteria.

The utility difference between the two concepts is called **regret**, and reflects the quality of the model, randomness of the environment, strategic behavior of the opponent and the cost of hedging against such strategic behavior. In the experiments presented in Sect. 5, the regret of different solution concepts is evaluated and compared.

The solution concept also tightly connects the security of the IDS system and the quality of the decisions it is able to achieve. The first two concepts – dominance and max-min rule – do not require any knowledge of opponent's plans intentions or goals, as they only consider the information about player's own decision function. On the other hand, reaching Nash equilibria (or analyzing the conditional dominance problem) requires that both players either interact over a longer period of time, or have at least some knowledge of opponent's utility function. Otherwise, they would not be able to identify the equilibria and can gain less profit (or rather more loss) than when playing max-min.

Under some circumstances, it might be even rational to disclose some information about the system to the attackers, in order to avoid the solutions which leave both players worse off. However, the practical implementation of this concept may be challenging, and our original intuition regarding the usefulness of the Nash equilibria as a solution concept was sceptical. This was to some extent

---

[4] Challenges are prerecorded sets of network traffic that are manually labeled as legitimate or malicious and can be seen as training samples.

disproved by the results of the experiments from Sect. 5, where it performs on par with other concepts even without explicit information transfer.

4   Game Integration for Runtime Reconfiguration

In this section, we will describe the integration of the game-theoretical model with the adaptation process of a particular IDS. This integration consists of several steps: dynamic parameter estimation in the system, game definition, game solution and integration of results back into the system.

There are two existing integration options addressing the problems from the opposite sides of the spectrum:

- *Off-line integration*, when the game is defined and solved analytically and the system parameters are configured according to game results [16]. This is the most traditional way of using the game theoretical methods, as their use ensures that the system parameters are set to force the adversary into the selection of less damaging (or more rational) strategies. The advantage of this approach is relatively easy solution identification and low technical difficulty, but the disadvantage is the fact that the game solutions identify the behavior that is advantageous on average, and do not reflect the dynamic changes of assumptions, threat characteristics and background traffic.
- *Direct on-line integration*, when the game uses presumed adversary actions in the observed network traffic to define the game. The game is being defined by the actual actions of real-world attackers executed against the monitored system. This approach addresses the problem with game definition relevance by using the actual attacks and traffic background to define the game at runtime. The game is then solved as an optimization problem, but with several drawbacks. Direct interaction between the adversary and the adaptation mechanism makes the system potentially vulnerable to attacks on machine learning and adaptation algorithms [3], making the whole IDS potentially less secure than without the use of game-theory driven adaptation.

Our approach, named *indirect online integration* [17] combines the above approaches and provides interesting security properties desirable for real-world deployment. The solution uses the concept of challenges to mix a controlled sample of legitimate and adversarial behavior with actually observed network traffic and is a compromise between the above approaches (see Fig. 1). In this case, the real traffic background (including any possible attacks) is used in conjunction with simulated hypothetical attacks within the system. These attacks are then mixed with the real traffic on IDS input and the system response to them is used as an input for game definition. The major advantage is higher robustness w.r.t strategic attacks on adaptation algorithms, and lower system configuration predictability by the adversary, as the simulation runs inside the system itself and its results can not be easily predicted by the attacker.

This approach offers the optimal mix of situation awareness and security against engineered inputs. In this case, we actually play against an abstract

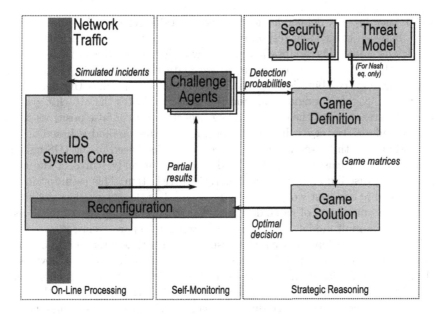

Fig. 1. Indirect online variant of integration of the game with IDS.

opponent model inside the system, and expect that the moves that are effective against this opponent will be as effective against the real attacks. The advantage of this approach is not only in its security, but also in better model character- istics in terms of strategy space coverage (less frequent, but critical attacks can be covered), robustness and relevance — the abstract game can represent the attacks and utility combinations that would be obvious only for insider attackers.

4.1 Indirect Online Integration

The use of the indirect online integration in practice requires a division of the covered time interval into sub-intervals defining each single game is a sequence. The length of such interval depends on the IDS technology used, line speed, hardware performance and other factors — it can vary between few seconds for pattern-matching packet filters to few minutes/hour for statistical anomaly detectors.

During the each interval t, the system measures/estimates the values of para- meters (in particular the detection probabilities $\alpha_{i,j}$ and the false positive proba- bilities β_i, V — discussed in details in Sect. 3.1). For the reasons listed above, we suggest the use of challenge-based parameter estimation [4], which relies on inser- tion of known instances of legitimate or malicious behavior into the background, unclassified traffic. We measure the system response to these challenges, drawn from the realistic attack classes, and use them to estimate the system response to all real-world samples from the same classes. In practice, we will define one class for each broadly defined attack/legitimate traffic type and measure the difference between the system response to legitimate traffic and to various classes of malicious traffic. The adaptation process will then assess the statistical properties of the response and use them to estimate the probability of detection of each strategy combination $\alpha_{i,j}$ (where the index i specifies the defender's strategy, i.e. system configuration, and the index j denotes the attack type, i.e. attacker's strategy) and the corresponding expected ratio of false positives β_i for given defender's strategy i. It is worth noting that this method is based on the assumption that the response of the detection method used in the IDS against members of each class is consistent and that the anomaly scores of the class members are distributed according to normal distribution. This assumption has been verified in our past work, and can be ensured as the attack class definition is under the full control of game designer — if the response to one of the classes is for example multimodal, it can be easily split into separate classes.

The game definition ordering with respect to each time interval also depends on the type of the underlying IDS. The CAMNEP system [18] is a NetFlow based collective anomaly detector, and therefore processes the data in well-defined and regularly produced batches rather than in real time — this means that the game is actually defined **after** the data has been recorded. In case of traditional pattern matching IDS that needs to operate on wire speed, the game needs to be defined and solved **beforehand**, so that the strategies can be applied directly to each processed packet, flow or connection. In practice, this means that the systems solving the game after the interval t on which the solution is applied have precise parameter estimations for each particular interval, while the wire-speed systems apply the t-th game results to the interval $t + 1$.[5]

In both cases, once the system obtains the game definition and solves it, it can directly apply the results back into the system configuration and use them on current or next time interval.

4.2 Game Strategies for Real World IDS

To test whether the game theoretical concepts can be integrated with a real IDS, we have used the CAMNEP system [18]. As we have noted in Sect. 4.1, the CAMNEP is a NetFlow-based IDS system. In addition, CAMNEP already features self-monitoring and self-optimizing functionality, allowing us to benchmark the performance of game-theoretical self-optimization with other approaches. The existing self-monitoring capabilities are also essential for online empirical estimation of the key utility function coefficients $\alpha_{i,j}$ and β_i (see Sect. 3.1), as their values typically evolve throughout the day.

The CAMNEP system is based on a self-organized, multi-level collaboration of detection agents, each of them maintaining an different model of traffic normality/anomaly. The agents share the anomaly estimates at various stages of

---

[5] The slight delay of application is unlikely to cause a problem, as suggested by our experimental results. The system using the parameters weighted over 5 last intervals performed comparably with the one using only the precise values for the specific interval.

processing and once they have reached their partial conclusions (anomaly scores for each network flow/connection), the system needs to aggregate these opinions together. At this stage, it is important to notice that the performance of individual detection agents and their combinations varies with background traffic and attack types. For more information about the CAMNEP system see Appendix A.

The defender strategies in CAMNEP are instantiated as specific aggregation functions used to integrate the opinions of detection agents in the system. Defender's strategy selection is thus technically straightforward, as it only picks one particular aggregation operator that aggregates diverse expert opinions with particular weights or methods. In our experimental system, there were 30 operators aggregating the opinions of 6 detection agents in total.

5 Experiments

The underlying CAMNEP system manages optimal selection of challenges and their mixing into the traffic. The selection of challenges is based on a simple threat model [4], which includes defender's risk estimates and potential losses. The response of the system to the challenges is then used both as an input for the original, trust-based self-adaptation mechanism and for the game-theoretical mechanism, running on the same traffic data and inserted challenges. This ensures that the experimental results, averaged over 40 system runs on the same inputs of the system fairly compare the influence of various techniques. The individual runs vary by the actually selected challenge values, as these are selected stochastically from a challenge DB. The data used for the experiments were acquired on a mid-size university network, with relatively low and stable background activity, and comprise of 100 5-min long intervals.

In our experiments, we want to measure two effects. First, we will compare the ability of the game-theoretical methods to deliver at least comparable performance on the inserted challenges. Then, we will judge the effectiveness of the selected configurations while detecting a classical, real world attack sequence comprising of exploratory activities: horizontal and vertical scanning, followed by password brute force attack on the SSH service on one of the vulnerable hosts. The results of the second experiment can be then used to measure how does the performance on challenge data translate to the performance on real attack detection.

We compare 6 different solution concepts for the defender: Trust1 and Trust5, MaxMin1 and MaxMin5 and Nash1 and Nash5. The name of the method depends on the strategy selection method, while the number (either 1 or 5) determines the number of periods over which the system behavior is observed: the concepts with the "1" suffix react to immediate situation only, while the solution concepts with the suffix "5" consider the values ($\alpha_{i,j}$ and β) aggregated over the last 5 intervals (25 min in total). In previous chapters we have described basic principles of the local self-adaptation along with the game-theoretical approach. Now we will discuss results measured on challenge-based attacks as well as results obtained on real-world attack.

In the first section we will describe basic settings of experiments such as state of network, types of performed attacks and settings of the utility function. Next, we will discuss results measured on challenge-based attacks with respect to concept called *regret minimalization* presented in [6] and in the third section we will compare these expected results with results obtained on real attacks. In the last section we will briefly discuss stability of each solution.

5.1   Experiments Settings

To correctly evaluate the system's ability to strategically select the best aggregation function we had to manually classify recorded data in order to obtain a dataset with a mix of partially classified third-party traffic and our attack with known properties. We have classified most of the legitimate traffic (roughly 75–80 % flows of legitimate traffic) and have manually performed attacks on the background of this traffic. The description of attacks we have performed is summarized in Table 1. From this table could be seen that we have simulated most common attacker whose goal is to gain administrator access to the protected system. Therefore, in these experiments are evaluated brute-force attacks with various speed and different types of scans used to discover vulnerable services such as horizontal scans, vertical scans, various types of fingerprinting, etc. The data used for the experiments were acquired on a mid-size university network, with relatively low and stable background activity, and comprise of 100 5-min long intervals.

To evaluate the concept of the game-theoretical approach we have used CAM-NEP which already implements basic process of the local self-adaptation along with managing number of challenges and its mixture with background traffic.

Table 1. Parameters of performed attack samples (real-world attacks).

Type of attack	Description
SSH bruteforce	Dictionary based attack with 100 attempts per attack passwords were randomly selected from predefined database
	Dictionary based attack with 300 attempts per attack, passwords were randomly selected from predefined database
Vertical scan	Vertical TCP scan performed against linux server with enabled OS detection
	Vertical UDP scan for all services performed against linux server
Horizontal scan	Horizontal scan for SSH service performed against network of Department of Cybernetics
	Horizontal UDP scan for DNS service
	Horizontal ICMP ping scan
	Combination horizontal and vertical scan performed against whole network of Department of Cybernetics

Table 2. Game parameter values for different strategies.

Attack strategy	$P_d = P_a$	γ_j
Horizontal scan	300	0.001
SSH brute force request	500	0.001
SSH brute force response	500	0.001
Vertical scan	300	0.001

In order to eliminate effects of random selection of challenges inserted into background traffic we had to perform 40 iteration on the same background data.

Next we have to specify all coefficients necessary to evaluate the utility function for attacker and defender as well (see Eqs. 4 and 5). Part of these coefficients is listed in Table 2. Note that the selection of the coefficients dependent upon the attacker's strategy corresponds with the coefficients obtained from attack trees. The rest of the coefficients are listed in Eqs. 17, 18, 19, 20, 21 and 22. Note that the selection represents situation when either defender nor attacker gain no asset (or worse in the case of attacker) when the attack is detected which covers attacks performed from Internet by unknown attacker.

$$C_a(a_j) = 0 \forall a_j, \tag{17}$$
$$C_{FP} = 1, \tag{18}$$
$$C_M = 0, \tag{19}$$
$$D_a(a_j) = 0, \tag{20}$$
$$D_d(a_j) - C_{TP} = 0 \; \forall a_j, \tag{21}$$
$$V(t) = 1 \; \forall t. \tag{22}$$

Finally, it is necessary to mention the solution concepts which are compared. We compare 6 different solution concepts used by defender: Trust1, Trust5, Max-Min1, MaxMin5, Nash1 and Nash5. The names of the solution concepts are derived from the name of the method (e.g.: Nash – Nash equilibrium, etc.) and the number of intervals used to evaluate the variables $\alpha_{i,j}$ and β_i. For example, the solution denoted by Nash1 refers to Nash equilibrium where is used no previous observation of the values $\alpha_{i,j}$ and β_i. On contrary, the solution named Trust5 refers to trust modeling which uses observation from last 5 intervals. Note that the *Dominated strategy* and *Conditional dominance* are not listed because the dominated strategy does not always provides results (the dominant strategy does not always exists) and, as it has turned out during the experimental evaluation, the conditional dominance return results comparable with stochastic selection of the best strategy and therefore it provides no measurable improvement to the detection process.

5.2   Challenge-Based Results

In this section we will present results measured on challenges artificially inserted into the background traffic. To fulfill this goal we have to define following evaluation function

$$\mathcal{E} = \frac{\bar{\theta}_y - \bar{\theta}_x}{\sigma_x + \sigma_y} \tag{23}$$

where $\bar{\theta}_y$ and $\bar{\theta}_x$ represent mean and σ_y and σ_x represent the standard deviation of legitimate and malicious challenges. This criteria rise when the detection process correctly separates the legitimate and malicious traffic (i.e. $\bar{\theta}_y$ tends to 1 and $\bar{\theta}_x$ tends to 0) and at the same time minimalizes their standard deviation which corresponds with the *trust experience* defined in [4]. Note that the challenges used for the following measurement equal to the challenges used to find optimal aggregation function using trust modeling approach – i.e. the training set. Therefore, in theory, the result of this experiment should show that the solution Trust1 should give the best results.

As could be seen in Fig. 2 this assumption has been partially confirmed. However, due to the fact, that the evaluation function does not completely refer to the trust experience used in the trust modeling approach, the results show that the Trust1 does not always provide the best possible result. Namely between the intervals 8–20 MaxMin1 outperforms Trust1.

But as could be seen on Fig. 3 and mainly on Fig. 2 the actual results confirms the theoretical assumption that the Trust1 and Trust5 outperform all other

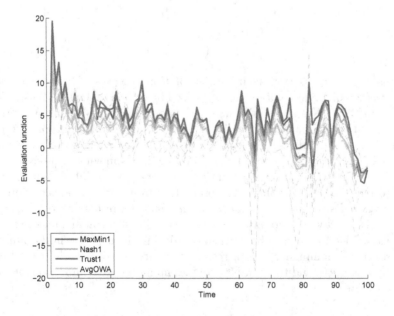

Fig. 2. Performance of solutions on challenge-based attacks over time, using the criteria 23 (higher is better) using values $\alpha_{i,j}$ and β_i without aggregation.

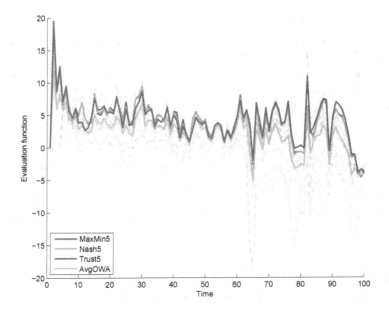

Fig. 3. Performance of solutions on challenge-based attacks over time, using the criteria 23 (higher is better) using values $\alpha_{i,j}$ and β_i aggregated over last 5 intervals.

Table 3. Average value of criteria 23 for challenge-based results

MaxMin1	Nash1	Trust1	MaxMin5	Nash5	Trust5	AvgOWA	Best
4.09	4.07	4.68	4.16	3.68	4.25	2.24	5.14

solution concepts. From the very same figures could be also seen that the rest of the solution concepts outperforms the average strategy even if it does not maximize the criteria used as evaluation function.

All results presented in Figs. 2 and 3 are summarized in Table 3 where the numbers represents average value of the evaluation function. From this table could be seen that all solution concepts highly outperforms the average strategy – *AvgOWA* (this value is computed as average value of evaluation function for all strategies in every time step). In the last column is shown average value for the best possible selection in every time step and could be seen that the all solution concepts provides relatively good results even in comparison to the best possible selection.

Finally, we have to mention that despite using solution concepts with short history provides better results on challenge-based attacks, solutions with longer history ensures more stable results and thereby provides higher protection against more sophisticated attacks.

5.3   Real-World Attacks

In the second phase of the experiments, we will verify the capability of the system to detect a real-world attack. To measure the quality of detection, we have to define a new criterion, since the one used for challenge insertion evaluation can not be applied to external attacks due to the lack of data. To solve this issue, we can use following criterion:

$$\mathcal{E}' = \frac{\bar{\theta}_{all} - \bar{\theta}_x}{\sigma_{all}} \tag{24}$$

where $\bar{\theta}_{all}$ represents mean and σ_{all} represents standard deviation of the whole dataset, and $\bar{\theta}_x$ represents mean of the all malicious flows (i.e. all flows with trustfulness lower than threshold). To evaluate the equality from the perspective of the false positives and false negatives we can define another criteria involving these two parameters:

$$\mathcal{E}'' = |FP| + 3|FN| \tag{25}$$

The variables $|FP|$ and $|FN|$ represents number of false positive and false negative events – i.e. the legitimate events labeled as attack and attacks labeled as legitimate traffic. The definition of this criteria implies that the lower value is better.

In the Table 4 are listed results for each type of performed attack. From this table could be seen that against the results measured on challenges, the differences between the results for individual solutions on the real-world attacks are reduced and in aggregated values even inverted. This fact is caused by the strategic behavior of the game-theoretical approach which is less dependent on the actually inserted challenges and rather follows the strategic behavior of the attacker than the actual performed attacks which is more suitable for the real assignment because in real situation there is no way to estimate the attacker's behavior. This conclusion is even further confirmed in Table 5 where are listed the results using criteria 25. This table shows that the solution Trust1 outperforms all others solutions at the cost of increasing the number of false positive and false negative events.

The results from the Tables 4 and 5 could be therefore summarized into the conclusion that the game-theoretical approach improves the results of the detection process and additionally provides more stability and robustness against adversary behavior of the attacker.

All discussed results are displayed on Figs. 4 and 5 for criteria $\mathcal{E}'$ and Figs. 6 and 7 for criteria $\mathcal{E}''$. Note that, the timing of performed attacks is listed in Table 6.

5.4   Solution Stability

In the last section we will briefly present the stability of each solution concept. We will compare all solutions noted in previous sections along with the *conditional dominance*. For each solution concept we will present two graphs – one when there is used no history and one where values $\alpha_{i,j}$ and β_i are aggregated over last 5 intervals.

Table 4. Results obtained on real world attacks using criteria 24

	MaxMin1	Nash1	Trust1	MaxMin5	Nash5	Trust5	AvgOWA	Best
SSH brute force	4.03	4.05	3.76	4.07	3.98	3.83	3.73	6.84
Vertical TCP scan with OS detection	0.86	0.81	0.79	0.94	0.95	0.90	0.79	1.47
Vertical UDP scan	2.44	2.26	2.43	2.40	2.37	2.37	2.20	2.95
Horizontal TCP scan for SSH service	0.65	−0.08	1.26	0.38	0.34	0.10	0.02	3.29
Horizontal UDP scan for DNS service	−1.49	−1.31	−1.48	−1.41	−1.41	−2.09	−1.77	0.23
Horizontal ICMP ping scan	4.30	4.27	4.78	4.79	4.84	4.37	3.89	8.62
Horizontal TCP scan for all services	3.10	3.11	3.01	2.99	3.00	2.98	3.21	4.51
Average	1.98	1.87	2.08	2.02	2.01	1.78	1.72	3.99

Table 5. Results obtained on real world attacks using criteria 25

	MaxMin1	Nash1	Trust1	MaxMin5	Nash5	Trust5	AvgOWA	Best
SSH brute force	23.56	23.31	21.86	20.46	20.05	21.01	23.35	19.24
Vertical TCP scan with OS detection	23.10	22.93	23.47	21.97	21.79	23.32	23.80	21.20
Vertical UDP scan	22.54	22.47	22.96	22.22	22.09	24.91	23.71	19.90
Horizontal TCP scan for SSH service	28.23	28.65	30.45	27.70	27.58	30.83	31.20	27.53
Horizontal UDP scan for DNS service	25.00	24.93	27.45	24.85	24.40	27.25	27.35	23.93
Horizontal ICMP ping scan	18.33	17.93	19.33	17.85	17.48	19.43	19.15	16.35
Horizontal TCP scan for all services	38.70	37.33	36.73	37.71	37.74	37.85	36.19	26.42
Average	25.63	25.36	26.04	24.68	24.44	26.37	26.39	22.08

Table 6. The timing of performed attacks

Attack	Interval
SSH brute force	14–17
Vertical TCP scan with OS detection	25–29
Vertical UDP scan	31–34
Horizontal TCP scan for SSH service	36
Horizontal UDP scan for DNS service	38
Horizontal ICMP ping scan	41
Horizontal TCP scan for all services	42–98

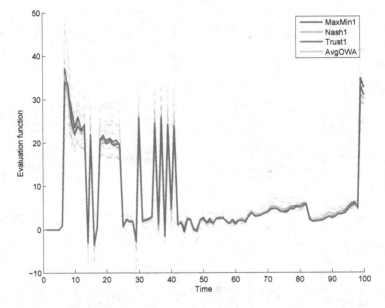

Fig. 4. Performance of solutions on real attacks over time, using the criteria $\mathcal{E}'$ (higher is better) using actual values $\alpha_{i,j}$ and β_i without aggregation.

At first we will discuss *conditional dominance*. Figures 8 and 9 show that almost all aggregation function have the same probability of usage during the whole experiment which implies that this concept does not reflect strategic behavior and provides almost the same results as random selection of the defender's strategy (these results were correctly confirmed on both challenge-based and real-world attacks). This is the main reason why we have not mentioned this solution concept in previous sections.

On the other hand, as could be seen on Figs. 10 and 11 the MaxMin solution concept provides relatively stable solution in both versions – with and without using history (Fig. 12).

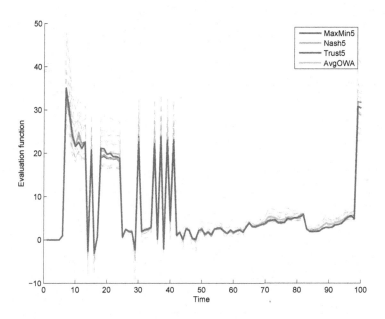

Fig. 5. Performance of solutions on real attacks over time, using the criteria $\mathcal{E}'$ (higher is better) using values $\alpha_{i,j}$ and β_i aggregated over last 5 intervals.

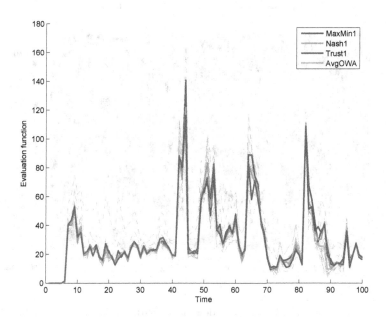

Fig. 6. Performance of solutions on real attacks over time, using the criteria $\mathcal{E}''$ (lower is better) using actual values $\alpha_{i,j}$ and β_i without aggregation.

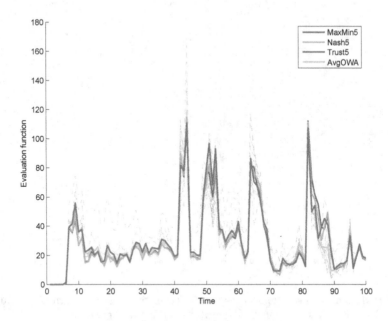

Fig. 7. Performance of solutions on real attacks over time, using the criteria $\mathcal{E}''$ (lower is better) using values $\alpha_{i,j}$ and β_i aggregated over last 5 intervals.

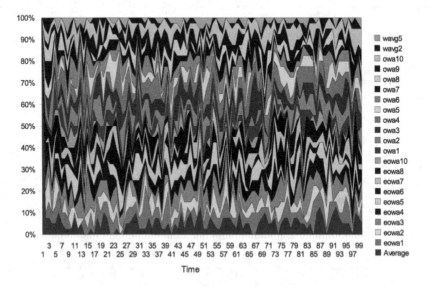

Fig. 8. Conditional dominance without history

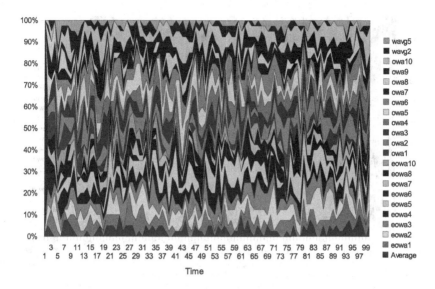

Fig. 9. Conditional dominance using history

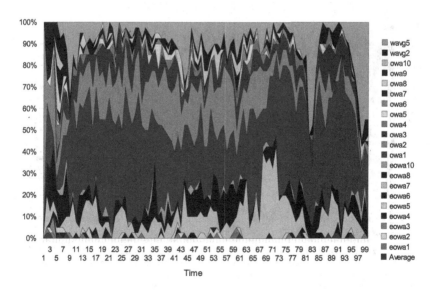

Fig. 10. Max-min rule without history

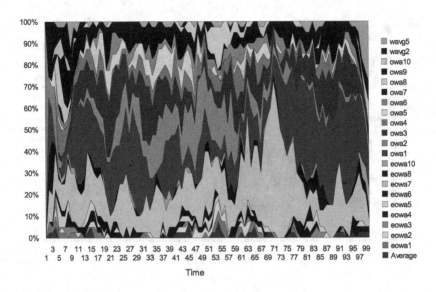

Fig. 11. Max-min rule using history

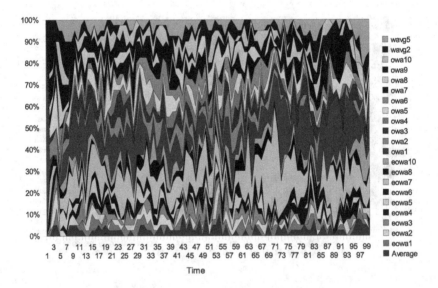

Fig. 12. Nash equilibrium without history

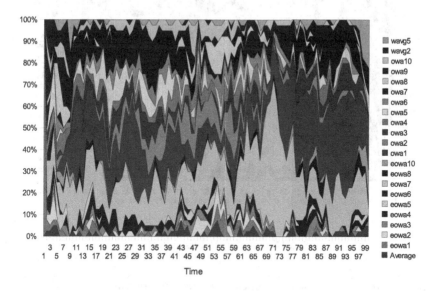

Fig. 13. Nash equilibrium using history

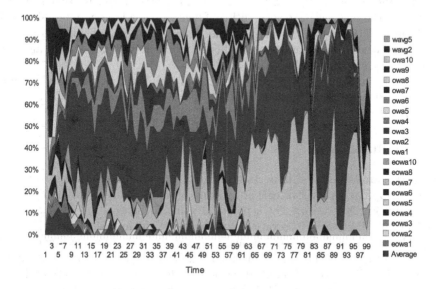

Fig. 14. Trust model without history

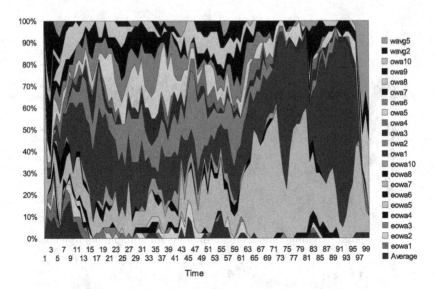

Fig. 15. Trust model using history

More interesting situation appears in the case of Nash equilibria. In the case that this solution concept does not use history, the selection of the best aggregation function is more stochastic and, as it was confirmed in Sect. 5.3, this fact affects negatively the final results. This phenomena is caused by the fact that without using history the utility function has a large number of mixed equilibria and the solver has to randomly select the best defender's strategy. In the case when the history is used, this phenomena does not appear and the solutions are relatively stable (Fig. 13).

Last solution concept is trust modeling approach which provides stable solutions in both versions – with or without history. The results can be seen on Figs. 14 and 15.

6   Conclusions

Our work addresses several important research issues related to the use of game-theoretic approaches for strategic adaptation of multi-agent intrusion detection system. In Sect. 3.1, we present a practical game theoretical model of the IDS problem and discuss its integration with a real-world intrusion detection system. The use of such mechanism shall improve system robustness w.r.t very advanced attacks based on adversarial machine learning approaches. The experiments performed with a simplified version of this commercially deployed IDS clearly showed that the cost of GT use is very low, and does not adversely affect the effectiveness of the adaptation process. In particular, our results suggest that the max-min method provides very consistent results, does not require an explicit model of opponent's utility function and is computationally trivial, making it an appropriate first choice for future implementations.

Besides the game theoretical model we present new possible way how to integrate game theoretical reasoning with the real-world IDS system. This is the main difference between our work and works presented in [2,7,11].

In our future work, we intend to perform actual experiments with targeted attacks on IDS. In this paper, we have made a first step - we show that multi-agent and game theoretical techniques do not harm the system - in the future, we need to show that they can measurably increase its robustness when facing highly sophisticated attackers, either human or software agents.

Acknowledgment. This material is based upon work supported by the ITC-A of the US Army under Contract No. W911NF-10-1-0070. Any opinions, findings and conclusions or recommendations expressed in this material are those of the author(s) and do not necessarily reflect the views of the ITC-A of the US Army. Also supported by Czech Ministry of Education grants 6840770038 and AMVIS-AnomalyNET. Also supported by MVČR Grant number VG2VS/242.

A CAMNEP

In order to evaluate the theoretical model in a production environment, we have used presented mechanism as a component of the CAMNEP network intrusion detection system [18], which is used to detect the attacks against computer networks by means of Network Behavior Analysis (NBA) techniques. This system processes NetFlow/IPFIX data provided by routers or other network equipment and uses this information to identify malicious traffic by means of collaborative, multi-algorithm anomaly detection. The system uses the multi-algorithm and multi-stage approach to optimize the error rate, while not compromising the performance of the system. The system contains two principal classes of classifying agents, which are able to evaluate the received traffic:

A.1 Detection Agents

Detection agents analyze raw network flows by their anomaly detection algorithms, exchange the anomalies between them and use the aggregated anomalies to build and update the long-term anomaly associated with the abstract traffic classes built by each agent. Each detection agent uses its own anomaly detection method, each works with a different traffic model based on a specific combination of aggregate traffic features. All detection agents map the same flows, together with the shared evaluation of these events, the aggregated immediate anomaly of these events determined by their anomaly detection algorithms, into the traffic clusters built using different features/metrics, thus building the aggregate anomaly hypothesis based on different premises. The *aggregated anomalies* associated with the individual traffic classes are built and maintained using the classic trust modeling techniques (not to be confused with the way trust is used in this work). The detection agents evaluate the anomaly of each network flow on the whole [0,1] interval, and the output of the detection agents is integrated by the aggregation agents.

A.2   Aggregation Agents

Aggregation agents α_i from the set $A = \{\alpha_1, \ldots, \alpha_g\}$ represent the various aggregation operators used to build the joint conclusion regarding the normality/anomaly of the flows from the individual opinions provided by the detection agents. Each agent uses a distinct averaging operator (based on order-weighted averaging or simple weighted averaging) to perform the $R^{g_{det}} \to R$ transformation from the g_{det}-dimensional space to a single real value, thus defining one composite system output that integrates the results of several detection agents. The aggregation agents also dynamically determine the threshold values used to transform the continuous aggregated anomaly value in the $[0, 1]$ interval into the crisp normal/anomalous assessment for each flow. The value of the threshold is either relative (i.e. leftmost part of the distribution) or absolute, based on the evaluation of the agent's response to challenges.

The detection and aggregation agents annotate the individual flows φ with a continuous *anomaly/normality* value in the $[0, 1]$ interval, with the value 1 corresponding to perfectly normal events and the value 0 to completely anomalous ones. This continuous anomaly value describes an agent's opinion regarding the anomaly of the event, and the agents apply adaptive or predefined thresholds to split the $[0, 1]$ interval into the normal and anomalous classes. The threshold used by the aggregation agents divides the flows into two classes: *normal* and *anomalous*. The anomalous flows are those whose anomaly falls below the threshold, while the normal flows are those whose anomaly is above the threshold. This distinction allows us to introduce the components of the error rate. *False Positives* (FP) are the legitimate flows classified as anomalous, while the *False Negatives* (FN) are the malicious flows classified as normal.

References

1. Kayacik, H.G., Zincir-Heywood, A.N.: Mimicry attacks demystified: what can attackers do to evade detection? In: Annual Conference on Privacy, Security and Trust, pp. 213–223 (2008)
2. Rubinstein, B.I.P., Nelson, B., Huang, L., Joseph, A.D., Lau, S., Taft, N., Tygar, J.D.: Evading anomaly detection through variance injection attacks on PCA. In: Lippmann, R., Kirda, E., Trachtenberg, A. (eds.) RAID 2008. LNCS, vol. 5230, pp. 394–395. Springer, Heidelberg (2008)
3. Barreno, M., Nelson, B., Sears, R., Joseph, A.D., Tygar, J.D.: Can machine learning be secure? In: ASIACCS '06: Proceedings of the 2006 ACM Symposium on Information, Computer and Communications Security, pp. 16–25. ACM, New York (2006)
4. Rehák, M., Staab, E., Fusenig, V., Pěchouček, M., Grill, M., Stiborek, J., Bartoš, K., Engel, T.: Runtime monitoring and dynamic reconfiguration for intrusion detection systems. In: Kirda, E., Jha, S., Balzarotti, D. (eds.) RAID 2009. LNCS, vol. 5758, pp. 61–80. Springer, Heidelberg (2009)
5. Nisan, N., Roughgarden, T., Tardos, E., Vazirani, V.V.: Algorithmic Game Theory. Cambridge University Press, New York (2007)

6. Blum, A., Mansour, Y.: Learning, regret minimization and equilibria. In: Nisan, N., Roughgarden, T., Tardos, E., Vazirani, V. (eds.) Algorithmic Game Theory, pp. 79–101. Cambridge University Press, New York (2007)

7. Alpcan, T., Başar, T.: A game theoretic approach to decision and analysis in network intrusion detection. In: Proceedings of the 42nd IEEE Conference on Decision and Control, Maui, HI, pp. 2595–2600, December 2003

8. Alpcan, T., Başar, T.: An intrusion detection game with limited observations. In: 12th International Symposium on Dynamic Games and Applications, Sophia Antipolis, France, July 2006

9. Liu, Y., Comaniciu, C., Man, H.: A bayesian game approach for intrusion detection in wireless ad hoc networks. In: GameNets '06: Proceeding from the 2006 Workshop on Game Theory for Communications and Networks, p. 4. ACM, New York (2006)

10. Chen, L., Leneutre, J.: A game theoretical framework on intrusion detection in heterogeneous networks. IEEE Trans. Inf. Forensics Secur. 4(2), 165–178 (2009)

11. Zhu, Q., Basar, T.: Dynamic policy-based IDS configuration. In: Joint 48th IEEE Conference on Decision and Control and 28th Chinese Control Conference, pp. 8600–8605 (2009)

12. Jain, M., Pita, J., Tambe, M., Ordónez, F., Paruchuri, P., Kraus, S.: Bayesian stackelberg games and their application for security at Los Angeles international airport. SIGecom Exch. 7(2), 1–3 (2008)

13. Becker, G.S.: Crime and punishment: an economic approach. J. Polit. Econ. 76(2), 169–217 (1968)

14. Ptacek, T.H., Newsham, T.N.: Insertion, evasion, and denial of service: eluding network intrusion detection. Technical report, Secure Networks Inc., Suite 330, 1201 5th Street S.W., Calgary, Alberta, Canada, T2R–0Y6 (1998)

15. Porter, R., Nudelman, E., Shoham, Y.: Simple search methods for finding a nash equilibrium. Games Econ. Behav. 63(2), 642–662 (2008)

16. Wagener, G., State, R., Dulaunoy, A., Engel, T.: Self adaptive high interaction honeypots driven by game theory. In: Guerraoui, R., Petit, F. (eds.) SSS 2009. LNCS, vol. 5873, pp. 741–755. Springer, Heidelberg (2009)

17. Rehak, M., Staab, E., Pechoucek, M., Stiborek, J., Grill, M., Bartos, K.: Dynamic information source selection for intrusion detection systems. In: Decker, K.S., Sichman, J.S., Sierra, C., Castelfranchi, C. (eds.) Proceedings of the 8th International Conference on Autonomous Agents and Multiagent Systems (AAMAS '09), IFAAMAS, pp. 1009–1016, May 2009

18. Rehák, M., Pechoucek, M., Grill, M., Stiborek, J., Bartoš, K., Celeda, P.: Adaptive multiagent system for network traffic monitoring. IEEE Intell. Syst. 24(3), 16–25 (2009)

+Cloud: A Virtual Organization of Multiagent System for Resource Allocation into a Cloud Computing Environment

Fernando De la Prieta[1]([⊠]), Sara Rodríguez[1], Javier Bajo[2],
and Juan M. Corchado[1]

[1] Department of Computer Science and Automation Control,
University of Salamanca, Plaza de la Merced s/n, 37007 Salamanca, Spain
{fer,srg,corchado}@usal.es
[2] Department of Artificial Intelligence, Technical University of Madrid,
Campus Montegancedo, Boadilla del Monte, 28660 Madrid, Spain
jbajo@fi.upm.es

Abstract. Nowadays Cloud Computing has gained in importance at a remarkable pace. The key characteristic of this technology is the possibility to provide new resources to the services in an elastic way according to current demand. In contrast to Cloud Computing, Multiagent Systems are focus on other features such as autonomy, decentralization, auto-organization, etc. This study demonstrates that this features of MAS are suitable to manage the physical infrastructure of a Cloud Computing environment, in other words, we present +Cloud which is a cloud platform managed by a Multiagent System.

Keywords: Cloud computing · Multi-agent system · Virtual organizations · Allocating resources

1 Introduction

The technology industry is presently making great strides in the development of the Cloud Computing paradigm. As a result, the number of both closed and open source platforms has been rapidly increasing. From an external point of view, the three most widely known services are Software, Platform and Infrastructure [5]. From an internal point of view, the services generally offered are considered elastic services. This means that it is possible to provide new resources to the services in an elastic way according to current demand. The main key factor of the rapid growth is that a high number of underlying technologies (virtualization, server farms, web services, web portals, etc.) which have reached their prime.

The reasons for the quick growth of the computational paradigm are varied, but it is possible to group them into three main categories. The first group is formed by the main technology companies (IBM, Google, Amazon, Microsoft, etc.) who have an economic interest in this paradigm as a new market. While previously an emergent market, its current dominance can be explained, in part, by a new business model which does not require an initial investment; instead, the client simply pays according

© Springer-Verlag Berlin Heidelberg 2014
R. Kowalczyk et al. (Eds.): TCCI XV, LNCS 8670, pp. 164–181, 2014.
DOI: 10.1007/978-3-662-44750-5_8

to the resources used (*pay-as-you-go* [7]) These companies oriented their efforts (economic, technological and human) to the development of this technology by creating various pilot projects (Sun Cloud by Sun, Blue Cloud by IBM, etc.), in addition to other open approaches [37] which eventually resulted in what we know as Cloud Computing. Secondly, the quick birth has been possible as a result of the maturity of the variety of technological components (server, cluster, high availability, grid computing) that form the computational paradigm; as well as the tremendous research, at both the hardware and software level, in incipient technologies such as virtualization [5]. Finally, there cannot be the slightest doubt that positive public reception has been a key factor in its rapid development. From the public's perspective, a cloud environment makes it possible not only to synchronize data, information, or even tasks, projects, etc., but also to work in a delocalized way through the use of online tools. And, from the companies' perspective, the main advantage is that a cloud environment does not require an initial investment, making it possible to pay only for those resources that are required at a particular moment.

Multi-agent System (MAS) have not played an important role in the development of the Cloud paradigm. According to Talia [38] it is possible to distinguish to groups: (i) MAS that use the computational features of a Cloud environment (processing, storage, etc.) [25, 32]; and (ii) Cloud environments that use MAS for the internal management of their resources, or to offer intelligent services. As is shown in the following section, the state-of-the-art indicates that the majority of the applications are related to the former group (agents using Clouds) [25, 33].

In short, Cloud Computing environment has a set of resources (physical and virtual) which have to vary dynamically in order to cope with the demand of the computational services being offered. MAS are suitable to help in the decision making about how vary dynamically these resources because its features (dynamicity, flexibility, autonomy, proactivity, learning, etc.) are exactly the features that a Cloud environment needs for the self-management of its resources. Within this model, the decision making process is complex, due to the variability of the demand for services and the lack of information on the decision components. This is the why an agent-based Cloud computing environment is suitable for the efficient allocation of computational resources, enabling the dynamic and automatic readaptation of each element which forms part of the cloud environment.

This study presents the +Cloud (masCloud) platform which is development by the BISITE Research group[1]. This platform allows to offer services at the PaaS (Platform as a Service) and SaaS (Software as a Service) levels. Both PaaS and SaaS layers are deployed using the internal resources of the cloud, in other words, the physical and virtual machines which provide a virtual hosting service with automatic scaling and functions for balancing workload. The core of this platform is a MAS based on a Virtual Organization (VO), which makes it possible to automatically manage the computational resources of the system, adapting them in an elastic way according to demand.

---

[1] http://bisite.usal.es/en

The present paper is structured as follows: the following section presents the state-of-the art of Cloud computing system and its relationship with MAS. The +Cloud architecture is then presented in detail. This study finalizes with a review of the initial tests of the system, conclusions and future research lines.

2   Cloud Computing and Multiagent Systems

Cloud Computing, understood as computational paradigm, is emerging recently with great importance. Although it may be initially considered another computational paradigm, reality indicates that its rapid progression is motivated by economic interests [7] in the underlying computational features.

Historically, the term *Cloud Computing* was first used by Professor Rammath [9]. However, the concept was becoming popular through Salesforce.com, a company that focused its market strategy to offer software as a service (SaaS) to big companies. However, IBM was the first company to detail the specific terms of the guidelines of this technology (auto-configuration, auto-monitorization, auto-optimization) in the document *Autonomic Computing Manifesto* [36]. By 2007, Google, IBM and others had joined together to form a research consortium which resulted in the birth of this technology as we know it today [26].

For the large companies, knowledge about this technology is a competitive advantage. First of all, the Cloud provider can offer its services through a pay-as-you-go model [7, 16], following the guidelines proposed by Utility computing [31]. Additionally, the Cloud user does not have to be concerned with demand peaks, transforming passive investments in operational expenses [4].

A large number of definitions [4, 7, 16, 28] have emerged at both a company and academic level. In each one, the authors try to highlight the most relevant features from their point of view. When a wide number of definitions are analyzed, it is possible to distinguish two big groups:

1. Those whose interests are focused on defining the technological aspects of the computational paradigm; these can be further divided in those who focus on defining either hardware or software characteristics.
2. Those whose interest is to highlight the aspects related to the negotiation model, which is intrinsically associated with a Cloud environment.

It was the American NIST (National Institute of Standards and Technology)[2] which defined Cloud Computing [5] as a *model for enabling ubiquitous, convenient, on-demand network access to a shared pool of configurable computing resources (e.g., networks, servers, storage, applications, and services) that can be rapidly provisioned and released with minimal management effort or service provider interaction.*

According to NIST, for a platform to offer Cloud Computing services, the services must contain the following characteristics [5]:

---

[2] http://www.nist.gov/

- Services on demand, meaning that services, regardless of their type, must be provided automatically and without human interaction according to user demand.
- Availability of services through the internet, meaning that clients should access the services through the internet and providers, as a result, must use this medium to provide their services.
- Availability of resources, meaning the provider must be able to offer services independently of their demand, using physical or virtual hardware resources assigned dynamically to each resource and reassigned according to demand. In this respect, there are authors such as [7, 43] who speak directly of high availability services, technology closely related to high availability computing.
- Elasticity, meaning that the different resources should be provided elastically and even automatically according to demand.

Further to this definition, NIST presents a set of features, different deployment models (private, public Community Cloud); and most importantly, three models of service. Understanding service as a capability that the Cloud offers to the end users, we can underscore the following three service models:

- Software as a Service (SaaS). This capability allows the provider to supply the user with applications that can be directly executed on the cloud infrastructure. This entails a number of advantages such as the ubiquity of the applications or the use of light clients. However, there are also a number of difficulties (which in some cases are strengths) directly related to the consumer's loss of control over the infrastructure (network, storage, operating system, difficulty to configure, etc.).
- Platform as a Service (PaaS). This capacity is supplied by the provider and allows the consumer to use the necessary tools to create their own applications within the Cloud environment. Some of these services include programming, libraries, tools, etc. As with the services in the previous level, the programmer does not control the underlying infrastructure, nor the environment where the applications are deployed.
- Infrastructure as a Service (IaaS); (o Hardware as a Service for Wang et al. [40]). This capability provides to the consumer include different kinds of hardware such as processing, storage, network, etc. This capacity can be provided to consumers with the ability to install their own software in an operating system deployed in a hardware environment, obviously virtualized, with characteristics defined by the actual user.

This division of models leads some authors to speak of Something as a Service (*ssS) [38].

2.1 Existing Platforms

SearchCloudComputing[3] provides a list of the 10 primary cloud computing providers, which include: VMWare, Microsoft Azure, Bluelock, Citrix, Joyent, Terremark y

---

[3] http://searchcloudcomputing.techtarget.com/photostory/2240149038/Top-10-cloud-providers-of-2012/1/Introduction

Amazon. Furthermore, from and academic and research scope, Corderiro et al. [13] propose the platform Euronet to interconnect multiple virtual laboratories using a Cloud environment. Given that the majority of Cloud platforms are proprietary and that the underlying infrastructure is invisible to researchers, Nurmi et al. [30] presents EUCALYPTUS, an open source framework for cloud computing that implements IaaS functions. Another proposal is provided in Malik et al. (2012), which presents a 3C model (Cooperative Cloud Computing) for research centers and universities. This model is based on the Virtual Cloud model and expects to generate a vast repository of computational resources for research centers.

All of these platforms have a very specific scope; that is, there is no platform that permits offering infrastructure, platform and software services in an integrated way. The scope of this study broaches this problem, facilitating the adoption of this paradigm by:

- developing the concept of multitenancy, which facilitates externalizing private and public Clouds to computing centers.
- developing elasticity models independent of the underlying technology, or very weakly coupled.
- developing an automated adaptation model for the internal hardware infrastructure, whether virtualized or not.

With regard to a negotiation model, cloud computing introduces a change in how to exploit and market a company's products. The model for acquiring hardware or software becomes a subscription model, or a service consumption model, which is essentially the same thing. This means that instead of acquiring pertinent computational resources, providers are hired instead. It is an attractive option for businesses, as it eliminates the requirements for the future planning of resources and permits beginning at the bottom and increasing resources only when they are actually needed. In their study, Artmbrust et al. [4] identify three case studies in which cloud computing is preferred to traditional storage: when demand for services varies over time, when it is not possible to foresee demand of services, and the computational efficiency that results from multiple machines.

2.2 Cloud and Agents

In a complex environment, such as that proposed in this project, it is difficult to determine when and how to carry out actions that imply changes in the operation or even the structure of the network. In the area of Distributed Artificial Intelligence, specifically in MAS technology, one of the goals is to create systems capable of making decisions in an autonomous and flexible way, and cooperating with other systems inside an organization. MAS technology is regarded as a potential technology to cope with the anticipated challenges of hybrid network operations. An analysis of the possibilities and benefits of implementing MAS shows that it is a suitable technology for the complex and highly dynamic operation of grid infrastructures, cloud computing, power systems or hybrid networks, among others [1, 21–24, 27].

Cloud environments require innovative architectures with advanced functionalities. To reach this objective, it is necessary to develop new functional architectures capable of providing adaptable and compatible frameworks, allowing access to services and applications regardless of location restrictions. A functional architecture defines the physical and logical structure of the components that make up a system, as well as the interactions between those components [17].

However, as indicated in the introduction, there are only a limited number of studies in the state of the art that relate Cloud Computing and agent technology [38]. In general terms, a Cloud system may use MAS applications in a Cloud environment for deployment, and there are also Cloud environments that use agent technology to manage their resources. Some of those applications include:

- **Agents using Cloud.** Within this group, the main state of the art applications use computational resources from the Cloud environment. For example, there are systems such as those described in [11, 15] that use the computational strength of the environment to perform simulations in different fields. Another example is presented in [10], where the Cloud environment is used as a persistence engine for information.
- **Cloud using Agents.** Within this subgroup, the range of possibilities is even further extended. Mong Sim [29] highlights three subgroups of applications: (i) combination of resources among Cloud providers; (ii) planning and coordination of shared resources; (iii) establishing contracts between users and Cloud service providers. As Mong Sim points out, it is possible to find studies such as [20, 39] that develop a Cloud service using agents for different specific purposes. Mong Sim used the Cloudle [29] which is an agent-based tool for discovering Cloud services. Some notable examples of Cloud providers combining resources include studies by Kaur Grewal [18] and Aarti Singh [35], which use shared Cloud resources to offer Infrastructure as a Service (IaaS) Examples that apply SLA to distribute services include [26, 31]. Finally we should point out the application of negotiation and agreement algorithms applied to different levels and processes within the framework of a cloud computing environment [19].

Recent tendencies have led to the use of Virtual Organizations (VOs), which can be considered as a set of individuals and institutions that need to coordinate resources and services across institutional boundaries. Therefore, a VO is an open system formed by the grouping and collaboration of heterogeneous entities; the separation between form and function that exists among them requires defining how a particular behaviour will take place. Multi-agent systems (MAS) technology, which allows forming dynamic agent organizations, is particularly well suited as a support for the development of these open systems. An open MAS organization modelling makes it possible to describe structural composition (i.e. roles, agent groups, interaction patterns, role relationships) and functional behaviour (i.e. agent tasks, plans or services), it can also incorporate normative regulations for controlling agent behaviour, dynamic entry/exit of components and dynamic formation of agent groups [3, 6, 8, 41, 42].

As a conclusion, the scope of this study is dealing with the open problem of re-source allocation over a Cloud computing paradigm. In this sense, one of the most appropriated approaches is to use MAS based on the VO, due to features such as of intelligence and adaptation that can enrich the capacities of a current Cloud Computing platform.

3   Proposed Architecture: +Cloud

+Cloud is a platform based on the cloud computing paradigm. This platform allows offering services at the PaaS and SaaS levels. The platform does not offer service at IaaS level. The internal layer is composed of the physical environment which allows the abstraction of resources shaped as virtual machines; however, the system does not offer this kind of service to the end users as shown in Fig. 1.

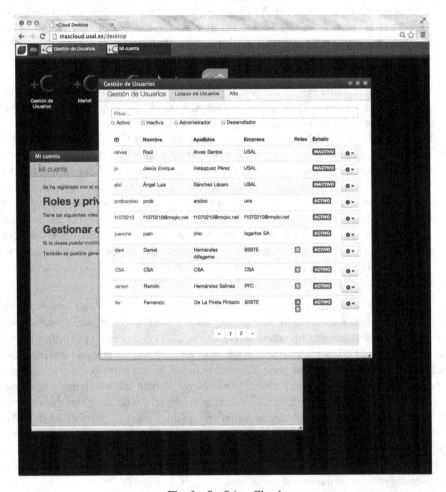

Fig. 1. SaaS in +Cloud

+Cloud has a layered structure that covers the main components of cloud computing:

- The SaaS layer is composed of the management applications for the environment (control of users, installed applications, etc.), and other more general third party applications that use the services provided by the next layer (PaaS).

  At this level, each user has a personalized virtual desktop from which they have access to their applications in the Cloud environment, and to a personally configured area as well. The virtual and physical resources are managed dynamically, but an overview of the internal resource can be seen through a specific web application.

- The PaaS layer provides services through REST web services in API format. One of the more notable services among the APIs is the identification of users and applications, a simple non-relational database service and a file storage area that controls versions and simulates a directory structure.

  The services of the Platform layer are presented in the form of stateless web services. The data format used for communication is JSON [14], which is more easily readable tan XML and includes enough expression capability for the present case. The existing service within the Cloud environment is:

  - The FSS (File Storage Service) provides an interface to a file container, emulating a directory-based structure, in which the files are stored with a set of metadata thus facilitating retrieval, indexing, searching, etc. The simulation of a directory structure allows application developers to interact with the service as they would with a physical file system. A simple mechanism for file versioning is provided. Web services are implemented using the web application framework Tornado[4] for Python.

  - The OSS (Object Storage Service) is a document-oriented and schemaless database service, which provides both ease of use and flexibility. In this context, a document is a set of keyword-value pairs where the values can also be documents (is a nested model), or references to other documents (with very weak integrity enforcement). Nevertheless, documents are not forced to share the same structure. A common usage pattern is to share a subset of attributes among the collection, as they represent entities of an application model. The allowed types of data are limited to the basic types present in JSON documents: strings, numbers, other documents and arrays of any of the previous types.

    As with the FSS, the web service is implemented using Python and the Tornado framework. By not managing file downloads nor uploads, there is no need to use the reverse proxy that manages them in every node.

  - The Identity Manager is in charge of offering authentication services to both customers and applications. The main capabilities of this service are (i) Single sign-on web authentication mechanism for users. This service allows the applications to check the identity of the users without implementing the authentication themselves; and (ii) REST calls to authenticate application/users and assign/obtain their roles in the applications within the Cloud.

---

[4] http://www.tornadoweb.org/

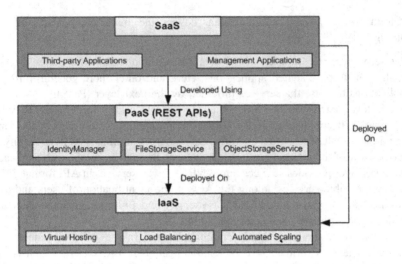

Fig. 2. +Cloud layer overview

The internal layer is used to deploy all management and general-purpose applications, in addition to the all services at the platform layer. This layer provides a virtual hosting service with automatic scaling and functions for balancing workload. It consists of a set of physical machines (servers) which contribute to the system by means of their computational resources (processing capacity, volatile memory, etc.). This level is formed by a large set of computational resources, referred to in previous technologies as a server cluster or server farm. Abstractions are performed over these hardware resources, as virtual machines, which allows the easy and dynamic management of computational resources. Although, performance decreases as a result of the computational needs of managing virtual computational resources, the advantages exceed the disadvantages, since complex tasks, such as the creation/destruction of virtual machines based on templates, the dynamic configuration of assigned resources, or even the migration of virtual machines between physical servers without stop the pending task, are made possible by virtualization.

In conclusion, a Cloud computing environment such as +Cloud platform can be viewed, at an external level, as a set of computational resources offered to end users. At an internal level, these services are deployed into a set of virtual machines that are hosted by physical server of the computational environment, as shown in Fig. 2.

The distribution of physical resources between the different virtual machines and between the different system services is a matter of current interest [19, 29, 35]. The redistribution of resources can be seen from both a micro and macro level point of view. From a micro point of view, there is a distribution of resources between the virtual machines that accommodate a single host. In other words, a physical server has a set of physical resources available (processing, memory and drive) that must be shared among the different virtual instances that it hosts, leaving a set of minimum resources available for its own host. At the macro level, there is a redistribution of resources at a global level in the Cloud, which entails migrating virtual machines in use

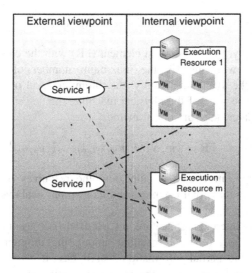

Fig. 3. Cloud computing deployment

between different servers, and turning on and off the physical machines that provide or consume resources within the Cloud environment.

The +Cloud platform uses a virtual organization of agents to manage the system resources. MAS can be perfectly adapted to solve this problem, as it allows making decisions in an open environment where the availability of information is limited and agents are thereby required to make decisions, amidst great uncertainty, that affect the entire system. As the decision making is a distributed process, the system has greater availability than other systems in which decision making is a centralized process.

Figure 3 provides a high level description of the system. As shown, the system is divided into the following agent organizations:

- **Resource Organization.** This agent organization is charge of managing both the physical and virtual system resources. The agents are distributed throughout the hardware elements of the Cloud environment. Their main goal is to maximize the use of resources. It is intended that there are no active resources that are underutilized, which implies that there must be the smallest possible number of active physical machines to satisfy the current demand. At the same time, the computational load of the active physical resources must be high. Within this organization includes the following roles:

  - *Local Resource Monitor*, in charge of knowing the usage level of the virtual resources of each virtual machine. There is one monitor for each physical machine and it has all the knowledge about the physical machine as well as its virtual machines.
  This agent keeps a vector with the information of the virtual machine (VR) hosted by the physical server, where each virtual machine is characterization by an id, kind of server, assigned computational capacity, assigned memory, state of the virtual machine, current use of the processing capacity and memory load.

$$VR = \{id, t, p, m, s_{VR}, u_p, u_m\}$$

In the same way, this role has an element (FR) with the characterization of the virtual server with the IP, the maximum number of virtual processors, the memory, the available memory, the minimum level of memory needed by the machine, the state of the physical machine, the percentage of current CPU use and the percentage of current memory use.

$$FR = \{ip, N, M, m_H, k_H, k_H, s_{FR}, u_p, u_m\}$$

– *Local Manager*, in charge of allocating the resources of a single physical machine among its virtual machines and its own physical machine. There is one in each physical server.
  In terms of its internal architecture, a CBR-BDI [12] agent can redistribute the resources of each physical machine among the different virtual machines according to the partial information that it has. It can modify (increase or decrease) the resources of the physical machines in use.
  Additionally, it can start up or shut down virtual machines within the local server; it does not do so by its own initiative, however, since it is the Global Regulator agent within the same machine that gives the order.
  The problem of redistribution of resources at micro level is the following:

$$Problem = \{\{FR, n_p, n_m\} | \{VR, n_p, n_m\}_1, \{VR, n_p, n_m\}_2, \ldots, \{VR, n_p, n_m\}_n\}$$

  Where n_p y n_m are the necessities of memory and processing capability by each virtual machine.
– *Global Regulator*, in charge of negotiating with its peers regarding the redistribution of the resources at a global level. There is one in each physical server. The Global Regular uses agreement algorithms between peers to distribute resources at a global level. When the service does not have the desired quality, all agents throughout the system with this role must reach an agreement as to how to solve the problem. To do so, they will use a distributed CBR system and algorithms according to [19]. Once the decision has been made, it is applied to the system in order to solve the problem.
– *Network Monitor*, this role can monitor the network from the point of view of each single physical machine. There is one in each physical server.
– *Hardware Manager*, the goal of this role is to manage at all times the hardware that is both in use and on standby. There is one in each physical server, each of which acts as coordinator.

• **Consumer Organization.** At the technological level this organization deploys over the computational resources offered by the organization described in the previous section. The services encompassed by this organization will, therefore use the system resources according to existing demand. Its main goal is to maximize the quality of service, which requires monitoring each service individually, keeping in mind that each service will be deployed simultaneously on various virtual machines located in different physical services.

- *Service supervisor,* this role is responsible for making decisions about each individual service. There is one for each service, each of which is located in the same virtual machine hosting the SDM of the same service, which in turn incorporates the load balancer service.
- *Service Demand Monitor,* in charge of monitoring each demand service which is offered by +Cloud. There is one agent of this type per each kind of service. They incorporate a load balancer to redirect requests to the different virtual machine which are offering the service at that time.
- *User,* represents the system users that use the services. As such, they are the ones that ultimately use the system resources. There can be different types of users: SaaS User, Cloud User and Administrator.

- **Management Organization.** This organization is in charge of ensuring that the entire system functions correctly, which is in fact its primary goal. There are two types of roles:

- *Global Supervisor,* which ensures that the other roles and agents of the VO work correctly. If something fails, or one of the agents does not respond to its messages, this role will take the necessary actions to restore the system to a functioning state.
- *Identity Manager,* is in charge of allowing other agents to enter or exit the system. Within the framework of this system, each time that a service is initiated or suspended, as with a physical machine, agents will enter or exit the system. This manager is also in charge of logging the User agents. As with all other services, this role, which itself constitutes a service, must be monitored as such (Fig. 4).

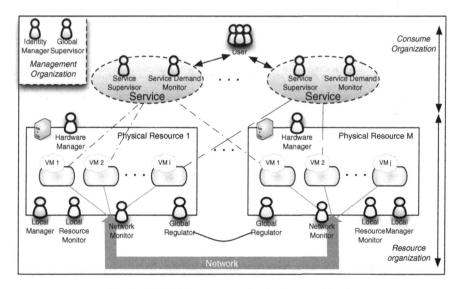

Fig. 4. MAS for resource redistribution in +Cloud

4 Preliminary Results

The +Cloud architecture has been continually evolving since its beginning. Beyond the multiagent system that governs the computational environment, it is also necessary to develop a set of services that are offered to the end user. Some of the more important services include FSS (*File Storage Service*), which stores files, and OSS (*Object Storage Service*), which stores information in a non-SQL database. The elasticity of these Cloud services is supported in the +Cloud architecture previously presented.

In order to obtain data regarding the performance of the Cloud environment and its ability to adapt to changes in the demand of services, a Denial of Service (DoS) attack is executed over the FSS. This is done by sending a constant stream of requests to the service over a period of 300 seconds. The number of requests is continually increased during the time of the test. This section presents the results obtained.

As indicated, the adaptation can be seen from both a micro and macro perspective. At the micro level, the adaptation takes place in each of the individual servers within the Cloud environment. This provides the physical server with limited capabilities (processing, memory, hard drive, etc.), which it must then share among the virtual machines it hosts. The adaptation takes place through the *Local Manager* agent, which works closely with the *Local Resource Monitor* agent. These agents are local to each machine and do not have information about the other nodes in the Cloud environment.

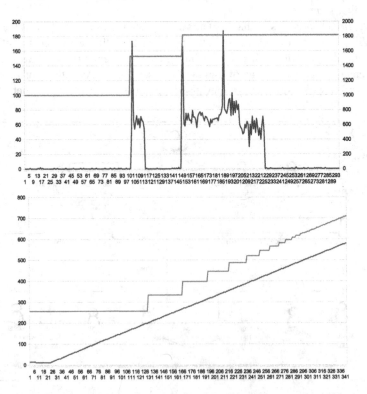

Fig. 5. Redistribution of resources at micro-level (upside: processor, downside: memory)

The *Local Manager* agent uses a CBR as a support system for making decisions. The information that this agent has is provided by the monitor, which is continually gathering data about the computational load of each machine in the physical server. Figure 5 shows the adaptation of the system. The graph on the upside shows an increase in the use of the processor in percentage (blue line) and the corresponding increase of CPU assigned by the Local Manager (red line). The graph on the downside shows the same process in terms of gigabytes, except that the memory is assigned to a virtual machine; the blue line is the memory used while the red line is the memory assigned.

When the *Local Manager* detects insufficient resources, or the *Service Supervisor* detects decreased quality of service, an adaptation process takes place at a macro level. This process, which can be reviewed in detail in [19], is based on the negotiation among the Global Regulator agents for different nodes in the Cloud environment. During this negotiation process, the agents decide how to redistribute resources among the different nodes (and not just internally) to rectify the problem of demand. Figure 6 shows how quality of service is improved. While the response time (y-axis) in the first part of service is very high, the quality of service improves considerably after two consecutive readaptations.

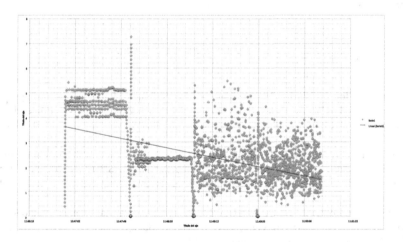

Fig. 6. Redistribution of resources at macro-level

5   Conclusions and Future Work

This study has presented the +Cloud platform, which is a Cloud Computing platform that is managed at internal level for a MAS based on a VO. As indicated in the introduction, initial results have shown that MAS technology is ideal for managing the computational resources of this type of system.

State of the art Cloud environments follow a centralized model for making decisions [37], which can lead to different problems such as (i) the need to centralize information; (ii) the need for a large computational load in the nodes where the decision making process takes place; and (iii) finally, the ability to recover from mistakes that can arise with a centralized decision making process. Using a model such as that proposed in this study, all of these problems can be resolved. This is due, in large part, to the system agents, which are able to make decisions in a Cloud environment where only partial information is available. Even if one or more nodes fail, it is still possible for the readaptation process to take place in the available nodes. Finally, the use of MAS allow using such techniques as the agreement techniques described in [19], which makes it possible to make decisions for readaptation at a global level without needing to centralize the information.

A final note with regard to future lines of work. Given the great technological component of the system and its dependency on the environment, we expect that future versions of the Cloud environment will include advanced concepts of MAS derived from the latest MAS methodologies, such as the concept of environment or rules to manage the actions of the roles within the organizations. The use of this kind of methodologies will facilitate the evolution of the platform and its independence from the underlying technological environment.

Acknowledgment. This work is supported by the Spanish government (MICINN) and European FEDER funds, project iHAS: Intelligent Social Computing for Human-Agent Societies (TIN2012-36586-C03-03).

References

1. Abras, S., Ploix, S., Pesty, P., Jacomino, P.: A multi-agent home automation system for power management. In: Cetto, J.A., Ferrier, J.-L., dias Pereira, J.M.C., Filipe, J. (eds.) Informatics in Control Automation and Robotics. LNCS, vol. 15, pp. 59–68. Springer, Heidelberg (2008)
2. An, B., Lesser, V., Irwin, D., Zin, M.: Automated negotiation with decommitment for dynamic resource allocation in cloud computing. In: AAMAS'10 Proceedings of the 9th International Conference on Autonomous Agents and Multiagent Systems, vol. 1, pp. 981–988 (2010)
3. Argente, E., Botti, V., Julian, V.: GORMAS: an organizational-oriented methodological guideline for open MAS. In: Gleizes, M.-P., Gomez-Sanz, J.J. (eds.) AOSE 2009. LNCS, vol. 6038, pp. 32–47. Springer, Heidelberg (2011)
4. Armbrust, M., Fox, A., Griffith, R., Joseph, A.D., Katz, R., Konwinski, A., Lee, G., Patterson, D., Rabkin, A., Stoica, I., Zaharia, M.: A view of cloud computing. Commun. ACM **53**(4), 50–58 (2010)
5. Barham, P., Dragovic, B., Fraser, K., Hand, S., Harris, T., Ho, A., Neugebauer, R., Pratt, I., Warfield, A.: Xend and the art of virtualization. In: ACM Symposium on Operating System Principles, Boltoin Landing, NY, USA, pp. 164–177 (2003)
6. Búrdalo, L., Terrasa, A., Julián, V., Zato, C., Rodríguez, S., Bajo, J., Corchado, J.M.: Improving the tracing system in PANGEA using the TRAMMAS model. In: Pavón, J., Duque-Méndez, N.D., Fuentes-Fernández, R. (eds.) IBERAMIA 2012. LNCS, vol. 7637, pp. 422–431. Springer, Heidelberg (2012)

7. Buyya, R.: Market-oriented cloud computing: vision, hype, and reality for delivering it services as computing utilities. In: 9th IEEE/ACM International Symposium on Cluster Computing and the Grid, 2009, CCGRID'09, pp. 5–13 (2009)
8. Carrascosa, C., Giret, A., Julian, V., Rebollo, M., Argente, E., Botti, V.: Service oriented MAS: an open architecture. In: Proceedings of the 8th International Conference on Autonomous Agents and Multiagent Systems, International Foundation for Autonomous Agents and Multiagent Systems, May 2009, vol. 2, pp. 1291–1292 (2009)
9. Chellappa, R.: Intermediaries in cloud-computing: a new computing paradigm. In: INFORMS, Cluster: Electronic Commerce, Dallas, Texas (1997)
10. Chen, C., Wang, K.: Cloud computing for agent-based urban transportation system. IEEE Intell. Syst. **26**, 73–79 (2011)
11. Cheng, Y., Low, M.Y.H., Zhou, S., Cai, W., Seng Choo, C.: Evolving agent-based simulations in the clouds. In: Third International Workshop on Advanced Computational Intelligence (IWACI), pp. 244–249 (2010)
12. Corchado, J.M., Pavón, J., Corchado, E.S., Castillo, L.F.: Development of CBR-BDI agents: a tourist guide application. In: Smith, I., Faltings, B.V. (eds.) EWCBR 1996. LNCS, vol. 1168, pp. 547–559. Springer, Heidelberg (2004)
13. Cordeiro, R.C., Fonseca, J.M., Donellan, A.: Euronet lab a cloud based laboratory environment. In: Global Engineering Education Conference (EDUCON), 2012, pp. 1–9. IEEE (2012)
14. Crockford, D.: The application/json media type for javascript object notation (json) (2006)
15. Dębski, R., Byrski, A., Kisiel-Dorohinicki, M.: Towards and agent-based augmented cloud. J. Telecommun. Inf. Technol. **6**, 16–22 (2012)
16. Erdogmus, H.: Cloud computing: does Nirvana hide behind the Nebula? IEEE Softw. **26**(2), 3–6 (2009)
17. Franklin, S., Graesser, A.: Is it an agent, or just a program? a taxonomy for autonomous agents. In: Jennings, N.R., Wooldridge, M.J., Müller, J.P. (eds.) ECAI-WS 1996 and ATAL 1996. LNCS, vol. 1193, pp. 21–35. Springer, Heidelberg (1997)
18. Grewa, R.K., Pateriya, P.K.: A rule-based approach for effective resource provisioning in hybrid cloud environment. Int. J. Comput. Sci. Inform. **1**, 101–106 (2012)
19. Heras, S., De la Prieta, F., Julian, V., Rodríguez, S., Botti, V., Bajo, J., Corchado, J.M.: Agreement technologies and their use in cloud computing environments. Prog. Artif. Intell. **1**(4), 277–290 (2012)
20. Jin Kim, M., Gun Yoon, H., Ku Lee, H.: IMAV: an intelligent multi-agent model based on cloud computing for resource virtualization. In: Lee, R. (ed.) Computers, Networks, Systems, and Industrial Engineering. Studies in Computational Intelligence, vol. 365, pp. 99–111. Springer, Heidelberg (2011)
21. Karnouskos, S., de Holanda, T.N.: Simulation of a smart grid city with software agents. In: Third UKSim European Symposium on Computer Modeling and Simulation, 2009, EMS'09, pp. 424–429 (2009)
22. Kok, J.K., Warmer, C.J., Kamphuis, I.G.. PowerMatcher: multiagent control in the electricity infrastructure. In: Proceedings of the Fourth International Joint Conference on Autonomous Agents and Multiagent Systems (AAMAS'05), pp. 75–82. ACM, New York (2005)
23. Lagorse, J., Paire, D., Miraoui, A.: A multi-agent system for energy management of distributed power sources. Renew. Energy **35**(1), 174–182 (2010)
24. Li, Y., Luo, Z.: A cooperative ad hoc routing based on cluster agent. In: IEEE Conference: 7th International Conference on Wireless Communications, Networking and Mobile Computing (WiCOM), Wuhan, People's Republic of China, pp. 23–25 (2011)

25. Li, Z., Chen, C., Wang, K.: Cloud computing for agent-based urban transportation systems. IEEE Intell. Syst. **26**(1), 73–79 (2011)
26. Lohr, S.: Google and IBM join in cloud computing research. New York Times (2007)
27. McArthur, S.D.J., Davidson, E.M., Catterson, V.M., Dimeas, A.L., Hatziargyriou, N.D., Ponci, F., Funabashi, T.: Multi-agent systems for power engineering applications—part II: technologies, standards, and tools for building multi-agent systems. IEEE Trans. Power Syst. **22**(4), 1753–1759 (2007)
28. Mell, P., Grance, T.: The NIST definition of cloud computing. In: NIST Special Publication, 800-145, pp. 1–3. NIST (2011)
29. Mong Sim, K.: Agent-based cloud commerce. In: IEEE International Conference on Industrial Engineering and Engineering Management, pp. 717–721 (2009)
30. Nurmi, D., Wolski, R., Grzegorczyk, C., Obertelli, G., Soman, S., Youseff, L., Zagorodnov, D.: The eucalyptus open-source cloud-computing system. In: 9th IEEE/ACM International Symposium on Cluster Computing and the Grid, 2009, CCGRID'09, May 2009, pp. 124–131 (2009)
31. Ross, J.W., Westerman, G.: Preparing for utility computing: the role of IT architecture and relationship management. IBM Syst. J. **43**(1), 5–19 (2004)
32. Schuldt, A., Hribernik, K., Gehrke, J.D., Thoben, K.D., Herzog, O.: Cloud computing for autonomous control in logistics. In: Ehrich, H.-D. (ed.) GI Jahrestagung. LNI, vol. 175, pp. 305–310. Springer, Heidelberg (2010)
33. Schuldt, A., Hribernik, K.A., Gehrke, J.D., Thoben, K.D., Herzog, O.: Cloud computing for autonomous control in logistics. In: 10th Annual Conference of the German Society for Computer Science (2010)
34. Siebenhaar, M., Nguyen, B., Lampe, U., Schuller, D., Steinmetz, R.: Concurrent negotiations in cloud-based systems. In: Vanmechelen, K., Altmann, J., Rana, O. (eds.) GECON 2011. LNCS, vol. 7150, pp. 17–31. Springer, Heidelberg (2012)
35. Singh, A., Malhotra, M.: Agent based framework for scalability in cloud computing. Int. J. Comput. Sci. Eng. Technol. (IJCSET) **3**, 41–45 (2012)
36. Stoica, F., Morris, D., Karger, M., Kaashoek, F., Balakrishnan, H.: Chord: a scalable peer-to-peer lookup service for internet applications. In: Proceedings of the Conference on Applications, Technologies, Architectures, and Protocols for Computer Communications, SIGCOMM'01, pp. 149–160 (2001)
37. Takato Endo, P., Estácio Gonçalves, G., Kelner, J., Sadok, D.: A survey on open-source cloud computing solutions. In: Brazilian Symposium on Computer Networks and Distributed Systems, pp. 3–16 (2010)
38. Talia, D.: Clouds meet agents: towards intelligent cloud services. IEEE Internet Comput. **16** (2), 78–81 (2012)
39. Venkataramana, K., Padmavathamma, M.: Agent-based approach for authentication in cloud. IRACST – Int. J. Comput. Sci. Inf. Technol. Secur. **2**(3), 598–603 (2012)
40. Wang, L., Tao, J., Kunze, M., Castellanos, A.C., Kramer, D., Karl, W.: Scientific cloud computing: early definition and experience. In: HPCC'08: 10th IEEE International Conference on High Performance Computing and Communications, pp. 825–830 (2008)
41. Zato, C., Sanchez, A., Villarrubia, G., Rodriguez, S., Corchado, J.M., Bajo, J.: Platform for building large-scale agent-based systems. In: 2012 IEEE Conference on Evolving and Adaptive Intelligent Systems (EAIS), pp. 69–73 (2012)

42. Zato, C., Villarrubia, G., Sánchez, A., Barri, I., Soler, E.R., del Viso, A.F., Sánchez, C.R., Cabo, J.A., Álamos, T., Sanz, J., Seco, J., Bajo, J., Corchado, J.M.: PANGEA - platform for automatic coNstruction of orGanizations of intElligent agents. In: Omatu, S., De Paz Santana, J.F., Rodríguez-González, S., Molina, J.M., Bernardos, A.M., Rodríguez, J.M.C. (eds.) Distributed Computing and Artificial Intelligence. Advances in Intelligent and Soft Computing, vol. 151, pp. 229–239. Springer, Heidelberg (2012)
43. Zhang, Q., Cheng, L., Boutaba, R.: Cloud computing: state-of-the-art and research challenges. J. Internet Serv. Appl. 1(1), 7–18 (2010)

Author Index

Printed in the United States
By Bookmasters